DEFENDING THE SOCIAL LICENCE OF FARMING

Issues, Challenges and New Directions for Agriculture

Editors: Jacqueline Williams and Paul Martin

Australian Centre for Agriculture and Law, University of New England

National Library of Australia Cataloguing-in-Publication entry

Defending the social licence of farming: issues, challenges and new directions for agriculture/edited by Jacqueline Williams and Paul Martin.

9780643101593 (pbk.)
9780643104549 (epdf)
9780643104556 (epub)

Includes bibliographical references and index.

Sustainable agriculture.
Agriculture – Environmental aspects.
Agriculture – Social aspects.
Agriculture and state.

Williams, Jacqueline.
Martin, Paul V.

338.18

Published by
CSIRO PUBLISHING
150 Oxford Street (PO Box 1139)
Collingwood VIC 3066
Australia

Telephone: +61 3 9662 7666
Local call: 1300 788 000 (Australia only)
Fax: +61 3 9662 7555
Email: publishing.sales@csiro.au
Web site: www.publish.csiro.au

Front cover: main image by Willem van Aken/CSIRO; other images by (left to right): CSIRO Plant Industry, Willem van Aken/CSIRO, CSIRO Plant Industry

Set in 10/12 Adobe Minion and Stone Sans
Cover and text design by James Kelly
Typeset by Desktop Concepts Pty Ltd, Melbourne
Printed in China by 1010 Printing International Ltd

CSIRO PUBLISHING publishes and distributes scientific, technical and health science books, magazines and journals from Australia to a worldwide audience and conducts these activities autonomously from the research activities of the Commonwealth Scientific and Industrial Research Organisation (CSIRO). The views expressed in this publication are those of the author(s) and do not necessarily represent those of, and should not be attributed to, the publisher or CSIRO.

Original print edition:
The paper this book is printed on is in accordance with the rules of the Forest Stewardship Council®. The FSC® promotes environmentally responsible, socially beneficial and economically viable management of the world's forests.

Contents

Preface

It is glib to say that society is becoming increasingly aware of the impacts that we are having upon the natural environment. What is less obvious is that this increasing awareness, borne of increasing competition for scarce resources (which encompasses demands for the conservation of these resources), is leading to fundamental changes in legal and social institutions. This includes marked changes to the practical meaning of property rights.

In the tradition of the philosopher Locke, property rights to the natural world and freedom to exploit the natural world were intertwined. In the mind of many farmers, and farming-linked political organisations, the dominant belief remains that with ownership comes a largely unfettered freedom to make full productive use of the resource, with few direct controls over the way in which the farm is managed.

However, in farming as in many other aspects of our lives, an individualist freedom paradigm is no longer a reliable guide to how to make your way in the world. Interdependence between potential users of scarce resources, realisation of the fundamental importance of 'environmental services', and the politics of the environment, are changing the nature of ownership and freedom to exploit. Increasingly, individual freedom is moderated by the requirements of interdependence, requiring that the citizen adjust their actions to ensure that the needs of the many are satisfied ahead of, or at least in conjunction with, the desires of the individual. In modern legal thinking, Locke's view of property and freedom to exploit has been substantially supplanted by a representation that considers the role of politics being at least as important as formal ownership in this relationship.

This is occurring at the same time as contemporary natural resource management places far more emphasis upon the use of private markets to achieve conservation and restoration of natural resources. These intersecting ideas, and resulting practices, result in a strange bifurcation of how property rights and natural resources are viewed together. On one hand, we see the proliferation of private rights to environmental assets, such as the creation of new tradeable private property rights to the extraction of water or biodiversity banking. On the other hand, we see increasingly stringent controls over the ways in which these and other private rights are exercised, as is illustrated by rules requiring the conservation of native vegetation on private farms.

Although an instinctive response to this change might be to seek to oppose it and to 'defend property rights', this is only one potential response. This traditionalist response in practice pits legalistic arguments arising from the Lockean tradition against contemporary views of the political boundaries of ownership. It also pits the interests of farmers against the interests of the broader community. In so doing, it translates some issues into a power-based argument, rather than seeking to find where the middle ground and fair compromise might lie.

An alternative response to the contest between private rights to exploit and public interests in conservation is to find ways to bring these two interests together. In this context, the idea of a farmer's *social licence* becomes as important as 'legal right' in determining the legitimate boundaries between freedom to exploit and obligations to conserve. Farmers who are able to demonstrate to society that their exploitation of natural resources does indeed maximise the benefit from their use and minimises the harms are in a far better position to gain the maximum freedom to benefit from their private property.

This book is intended to explore contemporary aspects of the *social licence to farm*, to provide those who are thoughtful about both the private interest and the public duty of farmers with insights into how this concept is developing in both theory and practice. However, lest readers think this is a concern limited to the farming sector, one only needs to consider recent debates about mining and other resource extractive industries to see a broader application. Such contests are demonstrated through the public media almost every day of the week.

Matthew Stevens in an article in the *Australian* newspaper (6 Nov, pp. 25, 29) entitled *New distrust of capitalism bad news for business* in commenting on the Canadian government quashing a takeover by BHP Billiton of the Potash Corporation of Saskatchewan, puts it thus:

> *'Across a host of the globe's most powerful and mature economies there has been a subtle lurch to re-regulation, provincialism and protectionism as communities and their political leadership seek responses to the failures at the root of the global financial crisis.*
>
> *We have emerged from the crisis with a new and elemental distrust of the pillars of capitalism and commercial globalisation and we are looking to governments and regulators, at the very least, to push the pause button on greed and adventurism.*
>
> *We have emerged too, in part as a product of this uncertainty and anger, with a collection of minority governments more vulnerable to fringe opinions fired by the new uncertainties.'*

One only has to consider the furore over private bank interest-rate setting, and proposals to re-regulate to control that private action, to see that social licence is a matter of strategic importance in almost every (if not every) industry.

This book documents the meaning of, and issues associated with, the social licence to farm. It provides examples of attempts to translate this concept into practice presented through case studies. We draw on a range of sustainability researchers with Australian and International experience who provide a much-needed critical review of the social licence dilemma. Critical analysis informs recommendations and guidance for land and water managers. This book is intended to assist sustainability policy makers, practitioners and the farming sector to assist them in navigating the challenging process of translating the social licence to farm from policy to practice in a new world economy.

As illustrated by examples in this book, it is possible to observe significant tensions between the farm sector, government and some parts of civil society in many countries. The romantic notion of farmers as the backbone of society, caring for the land and being valued by townsfolk for their contribution to society is gradually being replaced by images of farming as an industrial enterprise retaining little of its historical image as caring stewards. Stories of environmental harms from farming, community concern for risks of new farming technologies such as genetically modified organisms (GMOs), stories of animal welfare issues and the changing 'face' of farming as a large-scale and intense industrial activity all fuel changing community attitudes. At the same time, the community itself has been shifting its expectations of responsibility for the environment, animal and worker welfare, and the delivery of public-good outcomes from private activities of business. Community trust in the stewardship performance of farmers has often been eroded, and this has had significant practical impacts on farming as the community seeks to replace trust-based relationships with those that formally enforce increasingly stringent standards of stewardship and limit freedoms of farmers. Many farmers view these developments with alarm, because of the practical constraints that are imposed upon them and because they do not understand why their once assumed status as trusted citizens and stewards seems to have been eroded. Particularly for the small-scale 'journeyman farmer', practicing their craft in traditional ways and holding to traditional values (and not enjoying the economic benefits of large-scale industrial farming), this is a source of confusion and sometimes anger.

Though fuelled by public media and citizen activists, the tensions over values, activities and freedoms are 'played out' through the political system or the courts. Legislation and litigation are the instruments used to transform the relationship between farming and the community, and to force compliance with changed standards and expectations. The use of the law to impose change upon farmers is illustrated in many ways: legislation and litigation to change animal management to ensure new standards of welfare; the creation and the removal of 'right to farm' laws to protect farmers against certain civil actions or to protect farmland; the removal of financial supports for farming, or the creation of sustainability conditions for such supports; limitations to farmers use of public lands, such as for grazing; sustainability limits to water use; higher rates of pay and standards for farm workers; and, of course, new environmental regulations. For the farmer, these represent tangible proof of the erosion of their social licence to farm. They often do place restrictions on the freedom of the farmer to manage their property and affairs in ways that have been traditionally considered normal, and this can be seen as an erosion of legal property rights. Often the contests are characterised as being over property rights, but the reality is much more complex.

This book is intended to throw some light on what is happening to the farmers' social licence, standing back from the particular issues and attempting to see the pattern of change that is occurring. Different authors look at social licence issues in a number of countries and farming sectors to observe the dynamic of change, and the ways in which farmers and organisations charged with managing water and other resources are responding to these challenges.

The first three chapters establish the nature of the challenge of social licence, exploring the meaning and the dynamic of the social licence concept, some aspects of farmer's attitudes and expectations, and the way in which expectations of virtue in resource use have evolved and been applied to the farming sector. The following seven chapters explore the challenges that have emerged in many countries, spanning the north and south of the globe and most continents, but with a particular focus on the way in which these matters have evolved in Australia. These examples suggest that while some farm industries have been able to negotiate this changing world quite effectively, many others have not. There are identifiable characteristics of the farming bodies that have been able to cope better than others, and words such as 'leadership', 'voluntary accountability' and 'sound strategy' are likely to capture some of the reader's sense of what leads to success.

In the last seven chapters we look at the variety of legal and institutional strategies that are relevant to the issues raised in the preceding chapters. We consider the link between social licence issues and freedom of commerce in Europe, various aspects of voluntary reporting, innovations in legal arrangements (notably new duties of stewardship, and collaborative governance between farm organisations and government), and one concept for streamlining and integrating legal and institutional arrangements to better deal with the social licence issues of farming, in ways that reward virtue and reduce complexity. The final chapter seeks to draw together these strands, and suggest some directions for both scholarship and strategy on farmers' social licence.

This book started out with an intention of exploring what is happening with social licence and freedom to farm. We did not begin with the intention of advancing any particular 'solutions', but with the intention of providing a resource for those concerned with these issues. We have, however, arrived at some clear views. It is clear that effectively dealing with change in how society views farming depends on a purposeful and disciplined approach to rebuilding trust. This requires that the farm sector 'hears' and embraces unpalatable, and sometimes unfair, criticism. It requires that farmers have to embrace the challenge of changing farming, rather than trying to use tactical actions such as public reporting or public relations, or even legal fights to defend farming practices and freedoms, to avoid the painful necessity of adapting to changing community standards. Tactical actions are useful in support, but they are not sufficient to protect the social licence upon which the farm sector depends in so many ways.

Where farmers have taken a principled and adventurous approach to embracing increased accountability, it has been most effective when it has been married to a hard-nosed strategy that includes implementation of quite radical transformational change. Sloppy thinking about 'to who,' and 'for what' accountability is required is likely to result in ineffective and costly approaches.

It would be unfair to trivialise what responding to changed community expectations means to farmers. Responding in such a positive manner is no small 'ask' of farmers. It involves making significant investment and perhaps giving up opportunities for income. Transformational change of this scale and depth is of national significance. It requires heroic leadership, and real integrity in the face of difficult choices and conflict, and it offers much benefit to the larger community, the costs of which will be inevitably be borne by responsible farmers more than any other group. This book will, we think, demonstrate why change that may be painful and costly in the short term may sometimes be the most effective way to ensure continued support from the community, which, in turn, is necessary in so many ways to the viability of farming.

Acknowledgements

Much of the work that led to this book was conducted within CRC Irrigation Futures (2003–2010) and the authors acknowledge the opportunity the CRC provided to conduct grounded inter-disciplinary research.

About the contributors

Dr Andres Arnalds is Assistant Director of the Icelandic Soil Conservation Service. Andres has worked extensively on the development of community-based strategies for conservation and restoration of soil and vegetation in Iceland. He facilitated the successful landcare program 'Farmers heal the land' as well as other programs aimed at increasing farmers' involvement in caring for the land. Andres has a PhD in rangeland ecology and management from Colorado State University, USA.

Dr Claudia Baldwin is a regional planner with more than 30 years experience in policy, planning and management in the Queensland Government and Great Barrier Reef Marine Park Authority, as a consultant overseas and in Australia. She recently joined academia, teaching regional and urban planning at the University of the Sunshine Coast, and researching in social, institutional issues, governance and collaborative processes in water, coastal and land-use planning and management.

Professor John C. Becker is admitted to the bars of Pennsylvania and the US Supreme Court and is the Penn State University Professor of Agricultural Economics and Law. His prior involvements have been at Dickinson, the University of Arkansas, and the Drake University Schools of Law. His research spans social issues of farming (such as succession) and aspects of natural resource management, notably comparative water law and farmers' compliance with the law.

Professor Jürgen Bröhmer came to the University of New England in 2006. He has been the Head of the Law School since 2007. Jürgen studied law at Mannheim University in Germany, did postgraduate work on European Union law at Saarland University in Saarbrücken and received his doctorate in 1995 from Saarland University in Saarbrücken. His areas of expertise are European Union law, public international law, international human rights law and comparative law. He is an adjunct professor at Murdoch University in Western Australia, the Europa-Institute of Saarland University, Saarbrücken, Germany and the Faculty of Law at Union University in Belgrade, Serbia.

Dr Evan Christen is a Principal Research scientist with CSIRO Land and Water based at the Griffith Laboratory in the Murrumbidgee Irrigation Area. He is also an Adjunct Associate Professor with the University of New England. During his 19 years with CSIRO, his focus has been on the sustainability of irrigation, initially dealing with drainage and salinity control and the treatment and reuse of irrigation drainage and agricultural and municipal wastewaters. More recently he has focused on improving irrigation efficiency. In his work he has seen the increasing importance of the triple bottom line, initially focusing on the biophysical and economic dimensions. Through the Cooperative Research Centre for Irrigation Futures, he led the Sustainability Challenge project that also incorporated the social dimension in irrigation reporting.

Professor Donna Craig is an academic and specialist practitioner in the area of international and national environmental law and policy and has over 30 years experience in research, legal practice, teaching and working with communities, indigenous peoples' organisations, governments and corporations. Her publications emphasise the indigenous, social, cultural and human rights dimensions of environmental and social impact assessment and sustainable development. In private practice, Donna has developed corporate environmental management strategies, compliance training and environmental audit programs. Her current position is Professor of Environmental Law at University of Western Sydney. Donna is a Regional Governor of the International Council on Environmental Law and foundation and continuing member of the *Northern Territory Environmental Protection Authority* (2007–2010) and the Advisory Board of Greenland-based *International Training Centre of Indigenous Peoples* (ITCIP).

Mark Hamstead is a water policy consultant with 25 years experience in water resource management, largely in government agencies in Australia. Mark has particular expertise in water planning, water entitlements, water trading and development of policy and legislation.

Professor Michael Jeffery QC holds a Chair in Law at the University of Western Sydney, Australia and Heads the UWS Social and Environmental Research Group. He has been appointed to a number of key environmental law positions both overseas and within Australia over several decades including serving as Director of Macquarie University's Centre for Environmental Law, Deputy Chair of the IUCN's Commission on Environmental Law and Chair of the Environmental Assessment Board of the Province of Ontario (Canada). He currently serves as Deputy Chair of the NSW Environmental Defenders Office Board of Management. Michael has published and lectured extensively on a wide range of environmental and administrative law issues and is a recognised authority on international environmental law, biodiversity law, environmental impact assessment, climate change, environmental governance, public participation in environmental decision making and trade and environment.

Dr Amanda Kennedy is the Deputy Director of the Australian Centre for Agriculture and Law at the University of New England. Dr Kennedy has a background in commerce and a PhD in law. She is also admitted as a lawyer in New South Wales. Dr Kennedy has substantial experience in multidisciplinary research, spanning law, the social sciences and natural resources. She has conducted research on contractual aspects of natural resource management and institutional aspects of rural social policy. In recent times her research has considered ways to improve the effectiveness of laws affecting farmers, and of improving rural access to professional services.

Professor Paul Martin is the Director of the Australian Centre for Agriculture and Law at the University of New England. Paul has extensive experience in the law and in commerce, working closely with primary industry bodies. He has been a corporate lawyer and strategic consultant in the wine industry, horticulture and (in more recent times) as a researcher in water law and policy. He leads an international research team focused on the future design of natural resource management laws and institutions. He has an extensive list of publications and contributions to public policy, and is well recognised internationally for his scholarship on laws, institutions and the primary industry sector.

Wayne Meyer is the Professor of Natural Resource Science is the Director of the Landscape Futures Program as part of the Environment Institute at the University of Adelaide. He is a graduate of that University, with research experience in Texas and South Africa. His research career included 27 years in CSIRO as an irrigation scientist, systems modeller and sustainable agriculture research manager. His current research aims to help people identify new ways to use soils, water, vegetation and communities to develop more renewable landscapes.

Dr Andrew Monk has two decades of experience in auditing, certification and standards setting and commercial interests across the organic supply chain. Andrew has a PhD that focused on organic production systems and sustainability. He consults to public and private entities on environmental issues and management systems, while being managing director of an environmental sector services company. Andrew is a prior CEO and current director of Biological Farmers of Australia Co-op Ltd (BFA) and an adjunct associate professor at the University of New England, Armidale, NSW, School of Business, Economics and Public Policy.

Dr Luca Montanarella is the leader of the Soil Data and Information Systems (SOIL Action) activities of the European Commission Joint Research Centre in support to the EU Thematic Strategy for Soil Protection and the Common Agricultural Policy (CAP). His background is in agriculture and science, and has written many publications concerned particularly with the chemistry and structure of soils, and with issues that link soils policy and soils science.

Dr Guy Roth has comprehensive knowledge of the Australian cotton industry. He was Chief Executive Officer of the Australian Cotton CRC / Cotton Catchment Communities Cooperative Research Centre (CRC) for 4 years. He was previously a lecturer at the University of New England in the School of Rural Science and won a National Business and Higher Education collaboration award. Guy lives in Narrabri, NSW and, as a director of Roth Rural & Regional, Guy undertakes program management and specialised consultancy work for a range of clients, including: National Coordinator, National Program Sustainable Irrigation, Joint leader, Working Group, Primary Industries Standing Committee Water Use in Agriculture and National RD&E strategy. He is an Associate Fellow of the School of Rural Science and Natural Resources at the University of New England.

Dr Mark Shepheard is an Australian Endeavour Research Award recipient who completed his doctorate on the farmer's duty of care to the environment at the Australian Centre for Agriculture and Law (AgLaw), University of New England. Coming from a background working on sustainability and natural resource management research with CSIRO and with state government natural resource management agencies, Mark is interested in the practical implications for farming of legal and social duties. Mark holds a Graduate Diploma in Sustainable Agriculture from the University of Sydney and a Master of Science in Environmental Change and Management from the University of Oxford.

Christopher Stone lectures in the Centre for Environmental Law at Macquarie University's School of Law and undertakes research for the Centre for Agriculture and Law at the University of New England. His research focuses on using insights from the social sciences and other fields to understand the drivers and motivations behind sustainable behaviours, such as compliance with environmental regulations and corporate social responsibility.

Dr Vikki Uhlmann has worked for many years on sustainable water management with state and local governments and the private sector in Australia, as well as developing countries. With multi-disciplinary qualifications, her focus is on research and policy development, project management, community engagement and capacity building, and monitoring and evaluation – all with a focus on sustainability.

Dr Adrian Walsh is Associate Professor in Philosophy at the University of New England. He works mainly in political philosophy and applied ethics. He is an Associate Editor of the Journal of Applied Philosophy. In recent years he has worked mainly on questions concerning the philosophy of economics, the ethics of market activity and problems of philosophical method. He has written, among other things, on the commodification of sport, the problem of the just price and the use of thought experiments in ethics. He is working now on justice aspects of the allocation of water.

Dr Jacqueline Williams has been a post-doctoral research fellow at the Australian Centre for Agriculture and Law since 2007. She has over 20 years experience in natural resource management in Australia, having worked at local, regional, state and national levels encompassing forestry, catchment management and community-based natural resource management (NRM). Her areas of expertise include: regional NRM systems; NRM institutions and governance; translation and harmonisation of regional NRM to the property scale through property management systems; and NRM policy development and implementation in particular land use, biodiversity and water quality.

SCENE SETTING

1

What is meant by the social licence?

Paul Martin and Mark Shepheard

At agricultural shows and farm field days across Australia, petitions are being signed to defend farmers' property rights in the face of regulatory restrictions. The *right to farm* concerns raised by farmers include: controls on land clearing; mining exploration access; biodiversity protection; animal welfare controls and the failure of government to deliver water to satisfy legally secure extraction rights. Farmers point to many instances where legislative and administrative restrictions have greatly harmed individuals by limiting the ways in which they operate their farming enterprises, who have unwillingly borne the costs of satisfying the public desire to manage farm resources to achieve social and environmental purposes. Advocates of tight controls over farming point to counter examples where some farmers have harmed the environment, or acted in ways that violate widely received social expectations of responsible behaviour. These debates involve clashes of values, and wide divergences in perceptions of the facts.

The boundaries of farmers' freedoms are discussed on radio and TV, in Parliament and at rowdy meetings in front of Parliament. Media reports and political pundits selectively (and sometimes hysterically) feed the debates with conflicting opinions, as those involved struggle to win the political high ground that they believe will lead to their preferred position being reflected in laws, policies and administrative decisions. At issue is the degree to which owners of legal rights to land and water can fully use these resources to satisfy their economic needs, or (alternatively) the degree to which the government acting on behalf of society as a whole, can legitimately limit this private use.

Debates about where the boundary lies between private freedom and public control are of far more than academic interest. They affect the economic viability of farms and the strategies that can be used by government to pursue public interests. The benign sounding term *social licence* masks a heated reality of an evolving contest over land, freedom and the environment.

In this chapter we will discuss the concept of a social licence, and its implications for farming, and the tension between a rights-based view of property and a responsibilities perspective on freedom to use that property. We will use some examples to demonstrate that the issue of social licence is a vital practical concern for the farm sector, and how a failure to meet community expectations can result in significant economic losses to farmers. We will also expand a little upon the link between this concept, morality and responsibility, as a basis for the chapters that follow.

Social licence, politics and property rights

A number of factors are demonstrated by this conflict. The first is that, although the debates seem to be about legal interests, in practice property rights and politics are intertwined.

Property rights issues are complicated further by the adverse effects of messy administrative arrangements and the many regulations governing resource use. The second is that there is an unavoidable tension between freedoms to exploit what one owns and the power of society to restrict the exercise of these private rights or freedoms. The third is that there is little consensus about the values that ought be applied through the law and there is often a wide disparity of views about the facts. Claims that some people are being unfairly treated are countered by claims of facts that illustrate that these individuals are in large part the authors of their own harms, or that the harms they or their peers have created justify the treatment that they have received.

There are many ways in which society can restrict or expand the freedom of people within that society. The term *social licence to operate* is a shorthand way to describe the latitude that society allows to its citizens to exploit resources for their private purposes. The expression was coined by the oil industry (Mureau 2000) when Shell recognised as part of its strategic thinking that its commercial freedoms were limited by the licence that society provides it to carry out its business. This presaged a major shift in the emphasis on social acceptability and social engagement as ways to defend the commercial interests of the corporation. Although the term is not well defined (Gunningham and Sinclair 2004), it is clear what it connotes. If industry violates community expectations about how it ought operate, it is within the power of society to harm industry by a variety of means including legal constraints, market penalties such as consumer boycotts or, even in the more extreme cases, by direct, sometimes violent, action.

The farm sector faces similar challenges to other resource use industries such as mining or oil production. It requires access to natural resources, and the community has the power to reduce or place conditions on that access. For this reason social licence to operate is coming into sharp focus as a challenge of increasing practical significance to the farm sector as a whole as well as to individual enterprises.

Mureau (2000) points out that 'in agriculture, new ways of communication with consumers and society members are necessary to maintain the licence to produce in the future. Farmers and their organisations carry an important responsibility in implementing the changes society desires. The challenge is to design an agricultural sector, which lives up to the desires and expectations of consumers and society-members. The relationship between agriculture and society is multi-dimensional, complex and full of tensions'.

How society regulates private freedom

The law is the obvious mechanism through which society exercises formal power: to create and enforce private rights; to create and enforce public restrictions on these rights, or to place other controls on what citizens are legally permitted to do. Behind the formality of the law is the dynamic of society as it continually adjusts arrangements between citizens, and between citizens and the state, to meet evolving needs and changing circumstances. If society wants to change how farmers exercise the freedoms that they have through property rights, then society can (and does) change those freedoms. They can use the law to expand these rights (as has been done with the creation of water property rights) or to limit these rights, as is the case with ownership of carbon in trees or the granting of mining exploration rights over private land.

The current focus in farming circles on defending the sanctity of property rights, while an important issue in its own right, is somewhat misleading. Even when property interests seem to grant broad rights, in practice there are many ways in which society limits these rights. Other constraints are imposed on the farmer through laws to manage pollution, protect biodiversity, or protect animal or human welfare. These laws are complemented by administrative controls, controls imposed by the supply chain, consumer preferences, and boycotts, social pressure and even direct action.

Other citizens can also limit how owners can use their private property. In using your own property, there is a general obligation not to interfere with other people's rights to enjoy and exploit their property or infringe the security of their person. The nature of what is considered to be the protectable interest of a neighbour has shifted markedly over the centuries, in response to changes in societies' expectation. A good example of this is the increasing recognition by the courts and by local government of noise and odour nuisance that can arise from some forms of farming. Society has many formal and informal mechanisms to adjust the freedom of action of a property owner.

The question of social licence to operate is involved in many conflicts in the rural sector (Aalders 2002; Gunningham and Sinclair 2004) under the guise of duty of care, right to farm or property right concepts such as attenuation or takings (Hatfield Dodds 2004; Hone and Fraser 2004). It operates at the level of the individual farmer and for the farm sector overall. Whenever someone relies on the voluntary consent of another to carry on an activity, they can be said to have a licence (to operate). When the person who controls the resource is acting on behalf of themselves or other individuals acting in their own interests, that is a private licence. When that person is acting on behalf of the broader community in deciding whether an activity is to be permitted on behalf of society, then that is a social licence to operate.

Private licences to operate are well understood. Whenever you want to use something that another owns, you are seeking a licence from them. If I require your consent to graze my cattle on your land or to draw water from your dam, I am reliant on a private licence. The relationship between concepts of licence and concepts of property is close where law governs relations.

However, not all relationships are governed by the law. A strong person may prevent another from accessing resources even when they are not the owner of those resources. The consent of a bully with the power to stop you doing what you are legally entitled to can be considered a private licence, but without legal status. The relationship between power and licence is present regardless of law, because the existence of a legal right does not in practice mean that you necessarily have the capacity to exercise that right. The relationship between law, ownership, power and social licence to operate is a critical consideration for farmers (Raff 2003; Robertson 2003; Hone and Fraser 2004; Macintosh and Denniss 2004; National Farmers Federation 2004; Shine 2004; Raff and Cooke 2005; World Wide Fund for Nature 2005; Lyons *et al.* 2006).

The social licence to use water for farming

Nowhere is this more clearly evident than in relation to water use. The social licence of the irrigation sector is reflected in the volume and timing of water allocation, the security of water and other title, and the constraints under which irrigators and irrigation authorities operate. Such issues lie at the heart of the viability challenge for farming in many parts of Australia, as is so powerfully illustrated by the recent Murray–Darling Basin Plan furore. Only the issue of rainfall is more fundamental than legal entitlements for famers who need access to water. Individual farmers may have a tradeable entitlement to access water, and they may believe that this is a property right to water. They may share the belief with many other farmers that ownership of property includes a somehow sacrosanct right to exploit the resource within very broad limits (Hone and Fraser 2004; Macintosh and Denniss 2004; National Farmers Federation 2004; Murphy 2005). However, a tradeable entitlement cannot guarantee an allocation, and no property right is an un-attenuated right that carries an absolute entitlement to exploit (Mobbs and Moore 2002; Moss 2002; Robertson 2003; Raff and Cooke 2005). Society, through government, retains the right to determine:

- the extent to which it wishes to grant secure entitlements to water. This includes the determination of the extent to which such entitlements may be traded and the conditions under which they may be exercised.
- constraints that it wishes to place on property interests, including conditions on rights or regulated responsibilities, or prohibitions outside the property domain (such as environmental laws) which nonetheless effectively constrain the property interest.
- the administrative arrangements that govern the exercise of competing interests, particularly decisions about annual allocations and the extent to which they may be directed to the environment, irrigators or urban uses; or to particular parts of the irrigation system; or when and in what volumes they may be exercised.

Farmers, as a community and individually, are engaged in continuous negotiations with society (often through government) over the terms of their social licence. This negotiation may be for the purpose of creating new property entitlements, additional freedoms within the existing entitlements, or reduction of constraints.

Virtually any significant innovation in farming depends upon administrative decisions by government or private sector system managers. Again, this is well illustrated with water. The decision to adopt a new allocation regime, to change the computer models through which water flows are managed, to implement new pumping or monitoring technologies or to implement a new ownership or water use regime: all such significant decisions in most Australian irrigation systems are ultimately subject to approval (or at least benign acceptance) by an administering authority with a concern for public acceptance. Some specific examples from across Australia, which reflect the social licence challenges for farming serve to make this point:

- In the Sydney basin, as with all Australian peri-urban areas, one of the fundamental issues for primary producers is the extent to which they will be allocated fresh or reuse water, and the terms of that allocation (including the price). There is a continuous negotiation process about zonings and permissible uses of farmland, and frequent clashes over noise, smell and social amenity. The agencies that act on behalf of society in determining the social licence include the Premier's Department, the Department of Lands, local government, and the Department of Environment and Climate Change, (Department of Water and Energy, Sydney Catchment Authority and Sydney Water). One could readily argue that the reason why commercial farmers find it so hard to achieve arrangements that suit them is because the social justification for doing so is insufficient, relative to other interests that are seen as providing greater social benefit (Martin *et al.* 2008).
- In the Macintyre Brook in south-western Queensland, as in most of Australia, irrigators are under pressure because of administrative decisions about how water allocations will be made. They find themselves subject to regulatory/administrative arrangements that are complex and restrictive. In pursuing innovations to improve this, they are seeking a more liberal social licence. Considerations such as public acceptability and political ramifications can affect any such decisions (Martin *et al.* 2008).
- In the Limestone Coast region in South Australia, there is water use conflict between traditional horticulture and viticulture and the newer silvicultural enterprises. Part of this debate reflects the previously advantaged position that was given to forestry plantations through *Managed Investment Scheme* taxation incentives. This has now been substantially reduced, in part because of rural community and farm organisation protests against this special treatment.
- In the Colleambally region in New South Wales, water allocations are contested between towns and farms, between the environment and consumptive use, and between use today and use in the future. The social legitimacy and water use efficiency of the irrigation

sector are elements in these debates. Further, new enterprises have been proposed that would feed into this debate, particularly relating to biofuels production.

It would be easy to point to many other examples of the same type of political and legal dynamic without even considering water. The implementation of carbon-reduction strategies, genetically modified organisms, farm-safety rules, controls over imports, support for farm financial counselling and many other issues all demonstrate the strategic importance of the perceived legitimacy of the sector. In each instance, the claims of the farm sector are being contested relative to other interests. The questions that are being asked are about the contribution that the sector is making socially, economically and environmentally relative to other potential uses of the scarce resource. Farming has no choice but to engage in such debates if it is to retain or increase the social licence on which it depends. Issues of social licence are at the heart of the capacity of the farm sector to innovate and remain (or become) competitive.

Trust-building as a strategic imperative

Success in such politicised conflicts depends not only upon the substance of the (legitimately self-interested) case that can be made by farmers. It also depends upon the credibility of the sector, which is developed not at the moment of conflict, but over many years. Social trust is a key consideration in the maintenance of the social licence (Fukuyama 1995; Siegrist *et al.* 2005; Weber and Hemmelskamp 2005; Dovidio *et al.* 2006). During any heated political conflict, the arguments used by either side are discounted by the expectation of self-interest and weighed in the light of what is known about the social performance and the integrity of the sector. Perceptions of failures of responsibility by individual farmers, groups of farmers or farming institutions will undermine farming's credibility relative to other interest groups. The experience in recent times in the conflict over irrigation water demands and the environment (and the frequent complaints of irrigators that their interests have been insufficiently considered) might be taken to indicate that farming has not been as successful as it may wish in defending its social licence.

Trust is closely related to expectations. Embedded in these are standards of behaviour that go beyond the minimal expectations of regulation (Hutter 2005). We expect more of people than that they just comply with the law! Morals, ethics and values influence the development of expectations of responsibility (Epstein 1987; Bowie 1991; Carroll 1991; Moir 2001). Morality is concerned with the distinction between right and wrong. It is an internal matter not objectively measured outside the individual. Acting consistent with one's morals does not necessarily mean that others will judge your actions as good. Ethics has a broader meaning, and may include stated or unstated codes and rules of conduct for an individual and an organisation. Values are beliefs about what is important (the basis of valuation and evaluation) and moral values or corporate ethics are part but not the whole of these (Pearsall and Trumble 2001). However, for businesses to try to build their management decisions around satisfying other people's morals and ethics is difficult, if not impossible. Aesop's fable of the man, the boy and the donkey is ancient wise advice that in seeking to meet everyone's needs you are most likely to end up achieving satisfaction for no-one, including yourself.

Expectations of social responsibility and accountability

Responsibility is perhaps a better term than morality for discussing what is expected of farming as a sector and farmers as people, because it connotes a more tangible set of standards. Bovens (1998) has identified five literal senses for the word responsibility:

- responsibility as causality: 'You caused the issue and are therefore factually responsible for it.'
- responsibility as accountability: 'You are responsible for dealing with the issue even if you did not cause it.'
- responsibility as a capacity: 'You are a responsible adult.'
- responsibility as a role: 'Your job carries this list of responsibilities.'
- responsibility as virtue: 'You are a responsible person who acts in a virtuous way.'

Requiring that organisations act responsibly reflects the general expectation that an organisation is bound by duties of gratitude and citizenship to society and that responsibility comes with power and wealth. This debt to society view carries the expectation for business to act responsibly. This entails not causing avoidable harm, honouring stakeholder rights and adhering to the ordinary canons of justice (Bowie 1991). Increasingly, a dimension of expected responsibility is environmental stewardship (McKay 2006). Acting responsibly is more likely if decision makers expect to be held accountable to a particular norm of behaviour, through a formal institution or informal forums such as the media or customer feedback (Bovens 1998). Accountability can be enforced through legal means, such as civil action or policed implementation. It can come through direct economic action by owners or customers. It can come through loss of resource access, including loss of the social licence to use publicly controlled resource. It can also come through social sanction such as ridicule, shaming or social discomfort. The unstated social contract between the organisation and the community forms the basis for this ongoing relationship. However, as the terms of the contract are unspecified and constantly adjusting, it is a difficult contract to meet.

Four general categories of accountability for actions exist (Bovens 1998):

- transgression of a norm; i.e. harm and standard of care
- causal connection; i.e. reasonable foresee-ability
- blameworthiness; i.e. breach of duty
- relationship; i.e. proximity.

On this basis, accountability can extend to issues that may not give rise to a legally actionable duty. The standard legal tests of accountability based on proximity or causality, for example, may be looser in a social setting than in the courtroom. Public accountability and social licence are clearly closely linked concepts.

There are different views about the nature of the organisation, its ethical obligations, its operations and boundaries of responsibility (Longstaff 2000; Muller and Siebenhuner 2007). To operate in markets (in which there are winners and losers) or make management decisions that inevitably will hurt or offend someone, individuals or organisations make choices about their responsibility to others. These choices are necessary to function in a market place. Well-defined boundaries of responsibility ensure that the individual or the corporation does not become consumed by attempts to deal with an open-ended range of socio-economic and environmental ills. It reduces the risk that by trying to deal with everything, the corporation will be effective in doing nothing.

Developing a farm sector approach

Expectations of our responsibility to others are not fixed. What is expected in a street market in Bangkok or when trading horses in Pakistan is quite different from what is expected on the stock exchange in Berlin or Tokyo, or in transactions between family members. Expectations of the behaviour of an enterprise reflect the underlying values, socio-cultural norms and

The counter view

In his article 'The social responsibility of business is to increase profits', Nobel Prize winning economist Milton Friedman argued that the exclusive duty of business is to pursue profits and refrain from engaging in deception and fraud. He states 'discussions of the "social responsibilities of business" are notable for their analytical looseness and lack of rigor. What does it mean to say that "business" has responsibilities? Only people can have responsibilities'. His view is based partly on the belief that shareholders or other owners can themselves decide where they ought make their social good investments. He also states the executive 'would be spending someone else's money for a general social interest. Insofar as his actions in accord with his "social responsibility" reduce returns to stockholders, he is spending their money. Insofar as his actions raise the price to customers, he is spending the customers' money. Insofar as his actions lower the wages of some employees, he is spending their money'. Friedman argues that there may be an economic interest in devoting effort to community good, because this may generate goodwill and thereby profit.

Implicit in this argument is that managers who pursue other goals inevitably diffuse effort and focus, and that social goal pursuit can be a mask for inefficiency, self-interest and laziness.

'The social responsibility of business is to increase profits', *The New York Times Magazine* September 13, 1970.

perceptions in a society. This speaks of the need for organisations to have a sound process to make decisions about the goals of its social responsibility, the actions involved and the resource allocation that is required (Brooks 2005). This in turn suggests a search process (such as the process developed in the *Sustainability Challenge* research program, which is discussed later in this book) that involves relationships and dialogue to identify stakeholder expectations (Cramer 2005). Understanding expectations is only one part of the enquiry, because decisions have to be made about what of the expectations the enterprise ought, and is prepared to, meet. Unfortunately, to date, the specification of such processes remains at a very abstract level.

Three concepts: stakeholders, social contracts and legitimacy (Moir 2001) may help to narrow the enquiry as to whose expectations are relevant in shaping the social licence. A practical approach would involve establishing the significance of stakeholders based on the attributes of power, legitimacy and urgency (Carroll 1991; Moir 2001). Strategic leadership is critical to the development of strategies to protect the social licence (Lantos 2001; Cramer 2005) because potential accountabilities run across all levels and management functions (Stainer 2006). A strategy to protect social trust requires a sincere concern for public responsibility within the organisation's business philosophy, core values and principles (Lantos 2001). In practice, such a strategy involves a management approach incorporating goal establishment, planning, implementation, measurement, assessment and feedback (Stainer 2006). Leadership is likely to require a genuine dialogue with stakeholders as part of management, reflecting a genuine ethical intent (Muller and Siebenhuner 2007).

As a starting point, it is useful to identify what relationships are significant to organisational performance, and to understand the character and quality of those relationships (Longstaff 2000). Understanding stakeholders has been likened to building an alliance (Brooks 2005).

Alliance suggests an awareness of joint interests and a willingness to further those interests in a relationship. It involves accepting that working together has its limits, relationships will not always be harmonious and that learning from each other is essential (Brooks 2005).

This suggests that moves towards requiring triple bottom line (TBL) reporting may be plagued by inconsistency of measures, limited community confidence in the good faith of the managers involved, and wasted resources until such boundary setting that a methodology is produced and is widely accepted. Our work in this aspect is intended to provide greater guidance about where these boundaries might lie.

References and further reading

Aalders M (2002) 'Drivers and drawbacks: regulation and environmental risk management systems'. Centre for Analysis of Risk and Regulation, London School of Economics and Political Science, London.

Bovens M (1998) *The Quest for Responsibility: Accountability and Citizenship in Complex Organisations.* Cambridge University Press, Cambridge, UK.

Bowie N (1991) New directions in corporate social responsibility (moral pluralism and reciprocity). *Business Horizons* **34**(4), 10–56.

Brooks S (2005) Corporate social responsibility and strategic management. *Strategic Change* **14**, 401–411.

Carroll AB (1991) The pyramid of corporate social responsibility: toward the moral management of organisational stakeholders. *Business Horizons* **34**(4), 10–39.

Cramer J (2005) Company learning about corporate social responsibility. *Business Strategy and the Environment* **14**, 255–266.

Dovidio JF, Piliavin, JA, Schroeder DA and Penner LA (2006) *The Social Psychology of Prosocial Behaviour.* Lawrence Erlbaum Associates, Mahwah, New Jersey.

Epstein EM (1987) The corporate social policy process: beyond business ethics, corporate social responsibility and, corporate social responsiveness. *California Management Review* **XXIX**(3), 15–99.

Fukuyama F (1995) *Trust.* Penguin, London.

Gunningham N and Sinclair D (2004) 'Voluntary approaches to environmental protection: lessons from the mining and forestry sectors'. Australian National University, Canberra.

Hatfield Dodds S (2004) 'The catchment care principle: a practical approach to achieving equity, ecosystem integrity and sustainable resource use'. CSIRO, Canberra.

Hone P and Fraser I (2004) Extending the duty of care: resource management and liability. School Working Papers, Series 2004, Deakin University Melbourne.

Hutter B (2005) 'The attractions of risk-based regulation: Accounting for the emergence of risk ideas in regulation'. Centre for Analysis of Risk and Regulation, London School of Economics and Political Science, London.

Lantos GP (2001) The boundaries of strategic corporate social responsibility. *Journal of Consumer Marketing* **18**(7), 35–95.

Longstaff S (2000) Corporate social responsibility. *City Ethics* **40 (Winter)**.

Lyons K, Davies K, Cottrell E, Smagl A and Larson S (2006) The need to consider the administration of property rights and restrictions before creating them. In: *Adapting Rules for Sustainable Resource Use.* (Eds A Smajgl and S Larson) pp. 221–234**.** CSIRO Sustainable Ecosystems, Townsville.

Macintosh A and Denniss R (2004) 'Property rights and the environment – should farmers have a right to compensation?' The Australia Institute, Canberra.

McKay J (2006) Issues for CEO's of water utilities with implementation of Australia's water laws. *Journal of Contemporary Water Research and Education* **135**, 115–130.

Martin P, Williams JA and Stone C (2008) 'Transaction costs and water reform: the devils hiding in the details'. Technical Report 08/08. Cooperative Research Centre for Irrigation Futures, Richmond.

Mobbs C and Moore K (2002) 'Property: rights and responsibilities – current Australian thinking'. Land & Water Australia, Canberra.

Moir L (2001) What do we mean by corporate social responsibility? *Corporate Governance* **1**(2), 6–16.

Moss W (2002) 'Why the property rights debate is holding back reforms: a case for a focus on structural adjustment'. WWF Australia Working Paper, WWF Australia, Sydney.

Muller M and Siebenhuner B (2007) Policy instruments for sustainability oriented organisational learning. *Business Strategy and the Environment* **16** (3), 232–245.

Mureau N (2000) The concept of 'license to produce': definition and application to dairy farming in the Netherlands. In: *International Conference European Rural Policy at the Crossroads,* 29 June–1 July 2000, The Arkleton Centre for Rural Development Research King's College, University of Aberdeen, Scotland.

Murphy S (2005) *Farmers criticise native vegetation laws.* Landline Program, Australian Broadcasting Corporation, Brisbane.

Pearsall J and Trumble B (2001) *Oxford English Reference Dictionary.* Oxford University Press, Oxford.

National Farmers Federation (2004) 'Land and native vegetation policy resource security policy position'. National Farmers Federation, Canberra.

Raff M (2003) 'Private property and environmental responsibility. A comparative study of German real property law'. Kluwer Law International, The Hague.

Raff M and Cooke E (2005) Toward an ecologically sustainable property concept. In: *Modern Studies in Property Law – Volume III.* (Ed. E Cooke) pp. 63–88. Hart Publishing, Oxford, UK.

Robertson S (2003) 'Property rights, responsibility and reason'. A speech by the Minister for Natural Resources and Mines, Hon. Stephen Robertson MP, Queensland Government, Brisbane.

Shine C (2004) 'Using tax incentives to conserve and enhance biological and landscape diversity in Europe'. STRA-REP, United Nations Environment Program, Strasbourg.

Siegrist M, Keller C and Kier HAL (2005) A new look at the psychometric paradigm of perception of hazards. *Risk Analysis* **25** (1), 211–222.

Stainer L (2006) Performance management and corporate social responsibility. The strategic connection. *Strategic Change* **15**, 253–264.

Weber M and Hemmelskamp J (2005) *Towards Environmental Innovation Systems.* Springer, Berlin.

World Wildlife Fund (2005) 'WWF briefing – native vegetation regulation: financial impact and policy issues'. WWF Australia, Sydney.

2

Understanding the social obligations of farmers

Claudia Baldwin

Social licence acknowledges that the private sector has a shared responsibility with government, acting on behalf of society, to help facilitate the development of strong and sustainable communities. Similar to the mining industry, which needs to locate where the mineral is, farming is constrained to where soil, water, climate and transport conditions are favourable. Like the mining industry, farmers understand they need to get along not only with their neighbours, but meet broader societal expectations. A high social or environmental risk or unacceptability of their operation can affect enterprise viability, family livelihood and have consequences for a small community.

In general, farmers and their families live in the same place as their business. Their lifestyle is integrated with their work and surrounding environment. They are also part of a neighbourhood and community, often quite small in population, where individuals play an important role in supporting community well-being. This juxtaposes a usually competitive business or corporate life with a more cooperative community life.

This chapter draws on research in a case study about farmers' values about social and community obligations for sustainable water use in Australia. The complex interrelationship of environment, economic and social aspects of farmers' lives is used to illustrate views about public versus private interest. It reveals deeper values and beliefs about equity; that is, who benefits and who pays. This research highlights that farmers, like others, are complex individuals, are not a homogeneous group, and that sharing their understanding can contribute to finding an agreed approach to considering issues that are relevant to the preservation of their social licence.

The chapter also makes a further contribution in showing that a dialogue about values and perceptions can be conducted in many ways, not only through the conventional mechanisms of surveys, workshops and discussion. It illustrates one of the emerging techniques for engaging with the emotional connections and hidden meanings of landscapes and communities and, in so doing, does reveal some new aspects of farmer's views of what is important to them, and their responsibilities to other citizens and to the environment.

The case study

The case study is in a region in Australia that is economically dominated by irrigated agriculture, shows signs of stressed water resources due to over-extraction and has experienced conflict over water: the Lockyer valley in Queensland (Figure 2.1). The Lockyer is comprised primarily of small (100–1000 ha) horticultural farms growing food, turf and lucerne for dairy

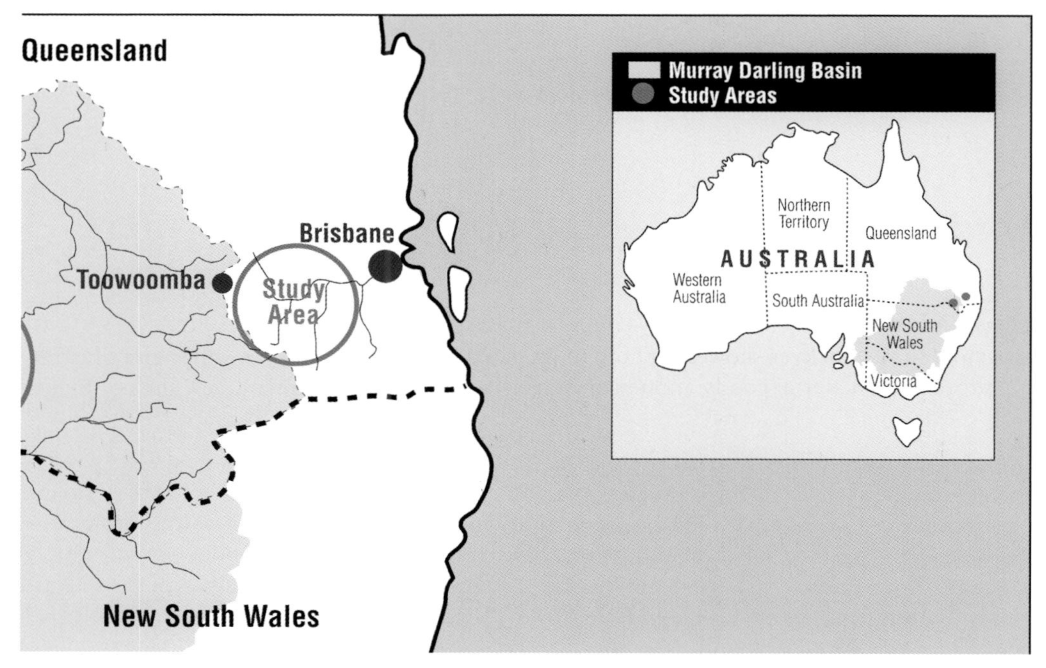

Figure 2.1: Location map of the Lockyer Valley

and racehorses. The variable subtropical climate and requirement for year-round cropping has resulted in an increasing drawdown of the groundwater aquifers to such an extent that there is often little surface water flow in this interconnected system, poor groundwater quality and decreasing or no yield from the groundwater bores in some areas. The geophysical characteristics mean that some irrigators have better access to reliable good quality groundwater and that pumping in one location can impact others.

Elsewhere in this book, it has been emphasised that the management of *social licence* issues intrinsically involves finding ways to have meaningful dialogues between farmers and others who might be impacted by farming activities. This dialogue must embrace issues of values and beliefs, and competing understanding of fact. Many of the common methods for such dialogue, such as discussion groups and workshops, or surveys, are well attuned to obtaining insights into factual aspects of the issues and stakeholder interests, but less so at exploring the more subjective and emotional components of the issues.

As outlined in the introduction to this book, and the first chapter, farmers' water access is mediated through administrative processes the outcomes of which to a large degree reflect the social licence that is given to farming, or forms of farming, relative to other uses of scarce water. Photovoice methodology was used in interviews and workshops to uncover the values and interests of 33 stakeholders faced with a state government water resource allocation planning process.

Participants were asked to take photos about what water meant to them and their community, and the meaning of sustainable water use, among other things. They were then interviewed individually using a semi-structured questionnaire with their photos as a 'prop'. They later shared and discussed the photos in a group workshop to derive an agreed direction. This was a slight adaptation of the Photovoice technique often used in community development projects for consolidating community views and empowering individuals (Wang and Burris 1997; Carlson *et al.* 2006).

This chapter focuses on the perceptions of the 21 interviewees in the Lockyer who use water in their agricultural operation, mainly through irrigation from groundwater, but some from surface water. Many of these were concerned that the government process might reduce water allocated to irrigation.

The identified priorities of farmers

Economic well-being

Consistent with a basic human need for economic well-being (Fisher *et al.* 1991) a majority of irrigators said that water meant they could earn a livelihood. As their first photo, 14 of the 21 irrigators in the Lockyer showed a photo of irrigated crops (Figure 2.2) or otherwise illustrating prosperity (material goods: money on a newspaper car advertisement, new shoes – Figure 2.3). Discussions of this representation triggered a number of considerations about the moral relationships involved in competition for water. One mentioned: '... you don't get too bloody greedy, then you might have some arguments'.

However, some irrigators expressed an alternative view, referring to the right to use water in spite of impacts on neighbours downstream: 'What's the moral obligation for a business-person to give his neighbour a profit? Is this communism or is it free enterprise?'

Others referred to 'self-interest':

> *'Selfishness applies to people that ... have a good supply of water ... As others ran out of water they were quite confident that they ... would command a high price for their produce. It hurts me to say this but I do honestly believe that some farmers do tend to be a bit selfish and short-sighted and not see the bigger picture.'*

Figure 2.2: 'Healthy crops ... it means money in my pocket and if I haven't got water I can't grow productive crops'.

Figure 2.3: 'Prosperity ... able to buy new shoes this year ... you have to make reasonable money to live'.

Public good

Whereas such statements reflect a focus on private interests, an indication of how irrigators perceive the domain of 'public interest' are reflected in discussion of distributional equity and the triple bottom line (TBL). To elicit values about distributional fairness, respondents were asked whether they agree or disagree with the following statements:

- Water is a community resource and should be managed for overall public good
- In allocating water, one should take account of impact on small towns and communities.
- Those upstream have a moral obligation to look after interests of those downstream.

Although there were slightly different interpretations of 'public good', 14 of 21 irrigators agreed that water should be managed for the 'public good'. Three irrigators provided insight into their rationale, suggesting that water used for irrigation produces food and supports the economy through taxes. Another queried whether it was in the public interest for water to be used to grow lucerne for racehorses: '... producing food in an efficient area to produce food, traditionally proven one of the best soils in the world ... so you get really good food, it's efficient because it's close to all the networks ... That's the best public good'.

It was noted that we need water to sustain our future and to sustain a supply of vegetables to the country: '... look at the public benefit ... One of the Lower Lockyer people figured that one meg water put on lettuce generates $1000 in tax'.

While almost all indicated that small towns and communities should have precedence for water, there was considerable disparity in views about allocation for the environment, and a concern about production missing out if environment was given a greater share. It was suggested that 'No one could disagree with the sentiment that the environment comes first but if you've got a person whose got three generations of surviving on the water he's got, it's a bit unfair to take it off that person'. In terms of looking after interests of those downstream, 15 of the 21 in the Lockyer responded positively: 'There should be a written law that they should look after those downstream'; 'Use shouldn't impact on your neighbour'.

Those who didn't agree were all from the upper catchment where there was more plentiful supply and their extraction could affect those downstream. Their rationale was: '... if you don't use it, you lose it'. This resonates with the earlier comment: 'What's the moral obligation for a business person to give his neighbour a profit'.

Sustainability

For effective management of water as a common pool resource, it is desirable to understand that cooperation is the way to achieve a mutual benefit for all (Slade and Levine 1986). Relevant considerations include: appropriate caring for the country; a stewardship approach; preventing environmental degradation; achieving a balance in the landscape and living in harmony with nature are ways to describe environmental values (Trumbo and O'Keefe 2005).

Three irrigators mentioned how their stewardship had improved the land – its productive capacity or the environment: 'We've just replanted it [the creek line] ... if you look after it you'll get good production'.

Another commented:

> *'... I can honestly ... take pride in saying ... there is not one bit of land that is not better now than it was when we bought it – we have left quite a legacy ... there's water underneath it and ... I'm pretty keen to protect it and use it as we need to ... we've been working over the years to get more trees into the area, to encourage more rain to happen, ... focused on maintaining the quality and quantity of water ... We've got four recharge holes where we help the recharge; we have a couple of hundred acres to revegetate where the salt water was coming out, we've allowed that to happen over the last 20 years by limiting stocking rates, keeping fires out of it, and not killing the suckers ... we're in a lucky spot to start with and we've been managing it ... fairly effectively ... The way trees need to be managed in the creek is different ... you'd keep that relatively clean, and then you allow trees to grow along the banks, to protect the banks and provide shade and all that sort of thing ... we can do what else we please, in an informed ... responsible way ... The proof is in the pudding ... when you look at the place and the fact that it's still got water.'*

Ten Lockyer irrigators felt that farmers need to use water efficiently, with one commenting: 'Farmers have got to manage water efficiently and if they are not prepared to do that they deserve to lose it'.

Many acknowledged that there should be a balance between environment and production. Others said it would be great to have water flowing in the Valley creeks but that this would affect livelihoods. Four irrigators, from Upper and Lower Lockyer, insisted that the environment received its share through rainfall into creeks and aquifers, or denied that it was an issue.

One aspect of environmental sustainability is intergenerational equity. Five irrigators referred to water conceptually in terms of water for future generations and long-term caring for the land or stewardship:

> *'I'm not here for a few years; I'm here for the long term. I wouldn't do anything to the vegies that I wouldn't let my kids eat ... When I was in high school I remember that creek being so deep that we could swim in it ... My kids will never get to experience that'.*

Overall, whereas an 'outsider' might have expected lack of flow in creeks to be a major sustainability and public benefit issue for the Lockyer, it was seldom mentioned (Figure 2.4). Instead the discussion was about benefits of more water in the aquifer to provide food for the nation and long-term use.

Figure 2.4: 'What wrong is there's no water in it. In January we had a bit of run in the creek ... the next day it was gone and within 4 days those water holes were all dried up'.

Sense of community

The research also examined sense of community, including consideration of measures that might show commitment to the community, length of residence; participation in community activities, a desire to relocate, how much participants like living in the community and how many people they know. Although specific photos were not referenced, it is noteworthy that a number of irrigators expressed their views in visual terms (e.g. 'look at the mountains', 'best view in the world', 'young people disappearing').

Most of the irrigators had lived in the region all their life, had local family ties and had no interest in relocating even if they could no longer farm:

> *'Where else do you want to live? ...You're not going to find a better place in the world to live, it's as simple as that. We can get in the back paddock and get a view that nobody else in the world can get. You look at the mountains. I go down the paddock and shift the irrigation and I got the best view in the world. Why would you want to do something else?'*

A sense of community was also conveyed by an irrigator who talked about how locals had helped him not long after he had moved in, when a storm was about to damage his crop. He said:

> *'People saw we were in trouble. We'd only been there for 6 months and three balers just appeared out of nowhere and two guys I still don't know who they were and as the last dozen bales rolled out, down it [the hail] came ... We love it; love the people.'*

Several farms severely affected through diminishing water resources had to reduce their workforce. The interrelationship of water with community viability is tangibly illustrated in Figure 2.5.

Figure 2.5: 'This is the car park out here where they park to go to the shed but as you can see it is empty because we don't have the work for these people. Believe it or not, I care for my workers and we don't have work for them. You try to give them 12 months work but you can't. It's very hard.'

In spite of such concerns, lifestyle was a main reason for living in the Lockyer, with many commenting that the clean environment contributed to their health and well-being. This is reflected in the following statements:

'You wouldn't do it for money ... you do it because you like it.'

'I wouldn't want any other job. I just like being out here in open air doing our thing growing vegetables.'

'I live in the country: clean air, a relaxed family lifestyle … I like being on the farm because … [I can] be home with my children.'

'There is stress here but a different stress [than working a nine to five office environment]… you're doing something you enjoy and are motivated about.'

Irrigators in all parts of the Lockyer found aspects of running a viable business stressful, especially with the lack of water, water reform and characteristics of the industry.

Discussion

The idea of a social licence tends to shape thinking along the lines that it is the community that directs farming activity towards sustainability and social justice. However, this is deceptive, because farmers themselves are generally concerned with the same issues. For them, the balance between public good and private interest is more complex than it is for those whose lifestyles and families will not be materially impacted by where this line is drawn. The need to create a living is an important element in how farmers view their social obligation, and their belief that they are indeed responsible stewards shapes their response to proposed further restrictions on their licence to manage the land in ways that they think are legitimate.

The concept of 'triple bottom line sustainability' seems to appropriately capture the farmers' thinking about their lifestyle where business, family and community are integrated with the land and a long-term view is required for business decisions, if nothing else. Ecologically sustainable development requires consideration of the wider economic, social and environmental implications in an integrated way, and a long-term rather than short-term view when taking decisions and actions.

Environmental and stewardship aspects of sustainability expressed by irrigators were primarily related to consumption: using water within the yield of the water resource and efficient water use. Although some commented on how they had improved the land, there was little practical interest in returning creeks to an environmental flow because the farmers recognised to do so would have economic impacts. The private interest was expressed in terms of long-term economic viability, cost of water and productivity from water. However, there was recognition that environmental and economic aspects were interrelated – if there was more water within the sustainable yield of the aquifer, there would be greater likelihood of surviving financially into the future. Water for production was also seen as a public good, because it provided food for the community and directly affected regional employment, population stability and community viability – from their perspective, contributing to 'public good'.

The farming lifestyle was interwoven with their community: economic and social roles were not divorced. The irrigators (with a couple of exceptions) clearly demonstrated a long-term commitment to each other and the Lockyer. By way of contrast, it is noteworthy that during the 3-year period of this research, half of the 12 non-irrigators interviewed had spent under 10 years in their location. Three years later, only one of the seven government interviewees are still in a position where they deal with the Lockyer. Yet irrigators pointed out several times that these are the people responsible for making decisions about their land and water.

Conclusion

There is a compelling case for the irrigation community to fully embrace 'triple bottom line' sustainability' as a goal to help secure and maintain their social licence to operate. In communities such as the Lockyer, it makes business sense to take responsibility for the state of the underground water on which irrigation depends. To maintain public benefits and community harmony, water needs to be used efficiently and there is a general obligation not to interfere with other people's rights or use. At this abstract level, there is a high degree of agreement among irrigators.

However, how to pursue this ambition in practice is more vexed than simply saying that it is a good idea. Based on the irrigator interviews, it was apparent that there was not a consensus on values across the Lockyer. To secure a social licence to produce, irrigators need to gain the trust and support of others that live and work in the area, and in particular those that make allocation decisions outside of the area. As has been illustrated in relation to organics, the value of the social licence can be greatly enhanced if the farm sector is able to rightly claim a moral licence, by showing that they genuinely and systematically manage their operations with a concern for the social aspirations of society, going beyond compliance and seeking to embrace consumer values such as sustainability and social justice, even when this comes at some cost to production. The feasibility of doing so is a natural concern of farmers.

A social licence is 'granted' by the community in general. Chapter 15 tells the story of how these irrigators came to a consensus on their future vision and their aim to take responsibility to 'co-manage' the resource. Their justification is based on perceived benefits, social obligations, and values about equity and independence. These are intangible. Unless effort is made to understand these beliefs and perceptions, and a process put in place to acknowledge and reconcile them, conflicts will continue around the issue of the social licence of farmers. Community relationships are dynamic, and at times fragile, and are not permanent, because beliefs, opinions and perceptions are subject to change as new information (such as a the state of the aquifer) is acquired or new factors (such as drought or water planning) are introduced. This makes managing for social licence a complex matter of continuous adoption, with choices continually having to be balanced against economic demands.

Acknowledgements

The author acknowledges the support of the Cooperative Research Centre for Irrigation Futures for the research on which this chapter is based.

References and further reading

Carlson E, Engebretson J and Chamberlain R (2006) Photovoice as a social process of critical consciousness. *Qualitative Health Research* **16**, 836–852.

Fisher R, Ury W and Patton B (1991) *Getting to Yes: Negotiating an Agreement without Giving In*. 2nd edn. Random House, Sydney.

Slade L and Levine H (1986) 'Water relations in foods'. Institute of Food Science Seminar, April 22, Cornell University, Ithaca, NY.

Trumbo C and O'Keefe G (2005) Intention to conserve water: environmental values, planned behaviour, and information effects: a comparison of three communities sharing a watershed'. *Society and Natural Resources* **14**, 889–899.

Wang C and Burris M (1997) Photovoice: concept, methodology, and use for participatory needs assessment. *Health Education and Behavior* **24**(3), 369–387.

3

The role of virtue in natural resource management

Adrian Walsh and Mark Shepheard

> *'They constantly try to escape*
> *From the darkness outside and within*
> *By dreaming of systems so perfect that no one will need to be good.'*
>
> *(TS Eliot 1934)*

In recent years in Australia, the most common policy response to a variety of urgent natural resource management dilemmas has been to rely on institutional mechanisms that appeal to private interest in the hope that these mechanisms will produce socially desirable outcomes. Market mechanisms have become the darlings of resource policy. This is clearly evident in debates over water policy, where a key assumption has been that the self-interest of market agents, as opposed to the good intentions of citizens, is the best way of managing this scarce resource. Behind this lies the thought that virtue is scarce and hence the good intentions of citizens cannot be relied upon; we should be thrifty in our dependence on virtue and, accordingly, our institutions should be *virtue parsimonious*.

Yet such tacit ideas sit oddly with the ideals of stewardship that are endorsed in other areas of government policy, in which there is an expectation that resource users, and in particular farmers, should be responsible and care intrinsically for the resources society has entrusted in them. Here, good intentions do seem to have a role to play. Policy on one hand is driven by a belief that farmers will adjust resource use only in pursuit of private property, and on the other hand that they ought be prepared to subordinate the same private interest to the broader public good.

In this chapter, we focus on the appeal to private interest as the *dominant engine* of public policy. We shall argue to the contrary that it is vital that we maintain ideals of responsibility and stewardship, which rely on the virtue of resource users and managers, as integral parts of policy. In making this case, we focus primarily on water use in Australia. We argue that minimising the reliance on virtue in water policy is not a rational response to the challenges facing us today. We shall defend policies that embed responsible use on the part of farmers and others into the core of our legal frameworks.

Virtue parsimony and institutional design in resource management

One striking feature of a great deal of public discourse about natural resource management, whether by farmers or government agencies, is the assumption that those who manage those resources should be motivated by ideals of sustainability and good environmental outcomes. They should be virtuous in the sense that they are motivated, in no small part, by a concern for the well-being of the environment and, in so doing, protect the interests of the community at large. Because they are given stewardship of the natural resources of the community, then they have an obligation to respect their community's interest in those resources.

This reasoning can be seen as part of an older, and more wide-ranging, philosophical tradition according to which, if we are to realise the 'Good Society', we require virtuous behaviour by citizens. Famously we see this idea in Plato's *Republic,* where it is the goodness of the citizens that collectively generates the goodness of the society as a whole. We will refer to any conception of the 'Good Society' that requires virtuous behaviour on the part of its citizens in order to function optimally (or even effectively) as a *virtue-rich* conception. The corollary is that if the members of a society are not so motivated then the society will not flourish.

If this virtue-rich ideal is an implicit element of much thinking in everyday discourse about natural resource management, then what is also striking is how distant it is from the basic assumptions of many of those engaged in designing institutions and policies for managing natural resources. According to many institutional designers, we cannot rely 'at all' on the virtue of resource users to realise good environmental outcomes, but instead must resort either to strong (and perhaps even Draconian) laws or to the use of market incentives. The fundamentals of these mechanisms is the pursuit of private gain, the public good being principally derived from wealth maximisation in the aggregate as a result. Here, the ideal is to minimise reliance on virtue in social policy formation and institutional design. Let us call this the tradition of *virtue parsimony.* This principle can be formulated as follows:

> *The Principle of Virtue Parsimony: If we are faced with a choice between two or more forms of institutional design, then, all other things being equal, we should prefer the one that requires less virtue input from its citizens to the one that requires more.*[1]

The leading thought is that it would be foolish to expect resource users to act with the social good in mind, so we must be parsimonious or thrifty with respect to our requirements of human goodwill.

Although there have been many who advocate the use of strong laws to prevent misuse of resources, in recent years, under the influence of the economic sciences, the focus has been on the use of the so-called *invisible hand* of the market to bring about desired outcomes. The market is a virtue parsimonious institution because it arguably makes use of little other than our desire to better ourselves financially.

At first sight, the motivation for this is somewhat perplexing. Why would one be inclined to endorse the suggestion that we should be thrifty in our reliance on the goodwill of resource users? In the virtue parsimony tradition, the reasons for doing so are two-fold: firstly, human beings are self-interested (and not governed by benevolent motives); and, secondly, given that fact about self-interest, public policy should function through appeals to self-interest alone, because it is more powerful than other motivations.

Foundational ideas

Let us begin with the scarcity of virtue. The idea that virtue is scarce is a touchstone in the work of many modern philosophers: one should assume that most people are knaves because they are primarily self-interested. As the 18th century Scottish philosopher David Hume (1741) notes:

'Political writers have established it as a maxim that, in contriving any system of government and fixing the several checks of the constitution, every man ought to be supposed a knave and to have no other end in all his actions than private interest.'

This is often paraphrased by economists into the language of scarce resources. Since the publication of Lionel Robbins' 1935 book, *The Nature and Significance of Economic Science*, economics has predominantly understood itself as being that discipline 'which studies human behaviour as a relationship between ends and scarce means which have alternative uses', (this is the so called scarcity definition of economics). We should economise on scarce resources. And one important resource economists have taken to be subject to such scarcity is human benevolence. Accordingly their attention is directed towards ways of organising our social institutions so as to 'economise on virtue'.[2]

The second move is to suggest that we should use this to society's advantage and govern by appealing to self-interest. As Hume (1741) notes: 'By this interest we must govern him, and, by means of it, make him, notwithstanding his insatiable avarice and ambition, cooperate to public good'.

Jeremy Bentham (Bowring 1843) makes a similar point about ensuring duty and interest coincide when he writes: 'Make it each man's interest to observe ... that conduct which it is his duty to observe'.

According to many modern thinkers we can achieve this aim of appealing to self-interest through the use of market-based incentives. In the market the self-interest of buyers and sellers bring about socially useful outcomes. As Adam Smith (1776) famously noted: 'It is not from the benevolence of the butcher, the brewer or the baker that we expect our dinner, but from their regard for their own self-interest'.

For Smith (1776), self-interest or self-love is causally efficacious in the production of certain material social benefits.[3] A similar point is made by the public choice theorist James Buchanan (1975), now talking of fruit-sellers instead of butchers and bakers:

'I do not know the fruit salesman personally, and I have no particular interest in his well-being. He reciprocates this attitude. I do not know, and have no need to know, whether he is in the direst poverty, extremely wealthy, or somewhere in between ... Yet the two of us are able to ... transact exchanges.'

Here the alleged scarcity of virtue is of little concern because of our ability to find mutually beneficial interactions.

In short, the point is that there is not much virtue to be found in the human realm; however, that is not cause for concern since pursuit of profit generates, via the 'invisible hand', socially beneficial outcomes. Therefore, we should refrain from developing policies or institutional mechanisms that require virtuous behaviour because to do so is both futile and unnecessary.

This is the kind of thinking that lies behind a great deal of recent natural resource policy making. The model of responsibility associated with water has shifted over time from a regime of private rights and morally inclusive obligations, to administrative enforcement of statute with its focus on accountability and punishment, and recently to a reliance on market mechanisms that reward the private good in defining the *most appropriate* access and use regime. Historically, shared responsibilities have played a more important part in managing water than they do now (Fisher 2004).

The historical notion of riparian rights is that the right holders along a natural watercourse are entitled to make an ordinary use of the resource without altering its quality or quantity

Table 3.1: Stages of water access and use in NSW

Pre 1912	Common law water rights
1912–70	Consumptive use and development aided by statute *Water Act* 1912 (NSW), *Irrigation Act* 1912 (NSW)
1970s	Emergence of pollution control concerns, for example *Clean Waters Act* 1970, and operation of the NSW State Pollution Control Commission from 1979 to 1991 (Replaced in 1991 by the NSW EPA)
1980s	Acknowledgement of the environment as a consumptive user through the *Water Administration Act* 1986 (NSW) Management concerns for water are broadened across agencies with the *Catchment Management Act* 1989
1990s	Emergence of sustainable river management in policy (NSW State Rivers and Estuaries Policy 1992) Principles of ESD incorporated into water administration statute. Corporatisation of state-run irrigation areas. Introduction of water trading as a mechanism to reallocate a scarce resource to higher value uses. Basin-wide cap on extractions in the Murray–Darling. Emerging national approaches to water reform through COAG.
2000	Introduction of *Water Management Act* 2000 (NSW): a centrepiece of water reform in NSW. Separation of river operations from planning, management and compliance functions.
2004	National Water Initiative. Facilitation of an efficient water market for trading within and between states and territories.
2007	*Water Act* 2007 (Cwlth), enacted in NSW law through the *Water (Commonwealth Powers) Act* 2008 (NSW) and amendments to other statute. Commonwealth buyback of water for environmental use commences
2010	Guide to the Murray–Darling Basin Plan.

(Howarth 1992). This means that each downstream riparian owner has the right to receive a flow of water unaltered in flow or quality and each upstream user has an obligation to ensure that is the case. Defining the access and use of water in this way focuses on moral obligations to protect the private rights of others, enforceable by civil litigation. A habit of caring for others is embedded in riparian rights.[4] The utility of these elegant and simple liability processes (an approximation of virtuous practice) has been lost in statute and administration. Table 3.1 uses the evaluation of water law and policy in the state of New South Wales to illustrate.

However, even the recent enforcement by market approaches, with its focus on the supremacy of private economic interest, lacks the potential to offer a plausible enforcement alternative to a reliance to at least some degree on ethical responsibility. We shall return to this point later in the chapter when reviewing the application of virtue ethics to our legal frameworks.

Why not virtue parsimony?

On what grounds might policy makers wish to reject the principle of virtue parsimony? The first point to make here is to recall that virtue parsimony is in considerable tension with a great deal of thinking about environmental issues: such thinking explicitly endorses ideals such as stewardship and responsible use. Stewardship here appeals to the virtue (by which we mean the good intentions) of resource users: it assumes them to be motivated by something other than greed and instrumental attitudes towards the environment. In this model, resource managers are concerned for the environment as an end in itself and have goals and motivations other than self-gratification.

One question we might ask is why ideals such as stewardship survive and continue to be endorsed as elements of social policy if virtue is as scarce as the virtue parsimony perspective would have us believe. Are those who endorse such ideals guilty of woolly romanticism about the human condition? We do not believe so. Although it is true that we cannot rely entirely on good intentions, human beings are not as vicious as writers like Hume (1741) contend. Humans are not knaves; instead we are beings capable of being motivated by the social good but who are often overly partial to our own interests.

Part of the problem is the construction of a false dichotomy between virtue rich social forms and virtue poor. If we reject the naïve idealism of those who think that society can be entirely run on goodwill, then it is assumed that we must endorse the view that we are knaves and hence adopt virtue parsimonious policy prescriptions. There is a hint of this dichotomy in the thinking of Alexander Hamilton (1788) in the *Federalist*: '… the assumption of universal venality in human nature is little less an error in political reasoning than the assumption of universal rectitude …'[5]

The dichotomy is false because it overlooks another position that we shall refer to as *political imperfectionism*. This is a weaker version of the claim about the self-interest of human nature according to which human beings are:

- self-interested, but not devoid of other-regarding motivations or concern for the common good
- subject to various kinds of temptation.

For the imperfectionist, the aim of political institutions is not to minimise the requirements of virtue, rather it is that we should make use of what virtue is available. Unlike the parsimony tradition, which has no threshold for the minimisation of virtue, political imperfectionism allows that people can often be motivated by virtue. It allows that even in pursuing their private interests they might be moderated or constrained by significant *other-regarding* concerns. According to the imperfectionist, we should attempt to make use of that virtue that is in fact on offer.

But why would any policy maker wish to do so? What reason is there to prefer to make use of that virtue that is available than minimise our reliance on it?

There are two related reasons. The first is a pragmatic point concerning the opportunities that any form of institutional design offers for agents to realise their self-interest in ways contrary to the aims of the design. All designs will inevitably have some flaw of this kind. If we take the market as our prime example, there will always be ways for commercial agents to make profits that are not in the interests of society as a while. There are cases where the 'invisible hand' does not work to bring about good outcomes and which economists typically refer to as forms of 'market failure'. We might think of them as examples of what some philosophers have called the 'invisible foot'.[6] Thus, in a commercial society, merchants might be able to realise more profits by substituting inferior components into their products. For instance, bakers might use a cheap chemical agent to minimise the amount of flour required.

No matter how well designed the institution, it is likely that there will always be such opportunities for agents to realise their interests in ways contrary to the common good. When such opportunities for self-advancement arise and there are no legal impediments, then all that prevents the occurrence of what might be a socially disastrous outcome is the goodwill and good intentions of the actors involved. We must rely on the virtue of those presented with such temptations. We have to rely on their commitment to the common good.

The point is clear in the case of the management of natural resources. Opportunities may arise for water users, for instance, to rort the allocations they are granted, in part because of the difficulty of policing water usage in remote locations. There will be incentives for them to profit and to over-use water in ways that are unsustainable for the system as a whole. Effective policing will be generally difficult, particularly if the 'cheat' is cunning and sophisticated. All that stands in the way of gross abuses here are the moral attitudes of the users: they need to be motivated in part by the health of the riverine system or respect for social values.

The second reason for developing systems that rely in part on – or have a clear role for – virtue is that if we develop such systems then we are more likely to foster virtue. If, as we have just argued, there will always be a need for virtue, then any system that fosters virtuous attitudes on the part of resources users is of great value. Equally, there are concerns that virtue parsimonious institutions undermine the moral attitudes we do have. If this is true, then this is further reason for us to encourage virtue richer forms of social design.

The latter line of reasoning derives in large part from the work of Albert Hirschman (1984) who opposes the idea of virtue parsimony and believes we require institutions that sponsor virtuous behaviour. He argues that sponsoring virtue parsimonious institutions undermines the virtue we do possess. In direct opposition to those who would maintain it is possible for people to quarantine market relations from one's general moral sentiments, which is sometimes knows as the 'separability thesis', Hirschman (1984) suggests that our promotion of institutions that idealise self-interest affects our *other-regarding* sympathies. His thought is that if we don't exercise these virtues, then they will not develop – a claim that Brennan and Hamlin (1995) mockingly refer to as the 'muscle-wastage thesis'.

There is an enormous empirical literature looking at these issues, the most notable of which defends what is known as the *Crowding-Out* thesis. According to the *Crowding-Out* thesis, as developed by the economist Bruno Frey (1997), market relations undermine our virtue. As Frey (1997) makes the point, assuming that we are knaves, virtue parsimonious policy (like payment for things normally treated as social obligations) crowds out what virtue we do possess. If we apply this to water resource management, the idea would be that developing laws and policies that assume that water users will always look after their own interests, rather than take in to account the interests of (say) the Murray–Darling Basin, will undermine whatever motives of stewardship and responsibility that those water users do in fact possess.

An even stronger line of argument than Hirshman (1984) is that virtues are necessary for the proper functioning of the market, and without them markets fail. This argument for the material necessity of virtue argues empirically that markets require virtues such as trust. Kenneth Arrow (1971) for instance, famously argued that:

> *'In the absence of trust ... opportunities for mutually beneficial cooperation would have to be foregone .... norms of social behaviour, including ethical and moral codes (may be) ... reactions of society to compensate for market failures.'*

There is a body of empirical literature exploring how *other-regarding* preferences and ethical commitments, such as trust and reciprocity, facilitate exchanges that in their absence would not occur and that losing those undermines markets.[7] The relevance of this to water resource management should be clear. If it is true that moral attitudes are required for a water market to function properly, then we should be wary of policies that attempt to minimise the reliance on virtue.

Be that as it may, the claim we defend here is the weaker one: that models of natural resource management that assume virtue to be a scarce resource will undermine the virtues we do have, while policies that involve some reliance on virtue are likely to foster it. Given the pragmatic point that any institutional design will provide actors with opportunities to act in socially detrimental ways in the pursuit of their own interests, we need resource users to have some concern with the social good. Hence any regulative system that encourages virtue is to be commended.

Virtue ethics and resource management

Thus far, we have been using the term virtue in a very broad sense to refer to other-regarding action; that is, to action directed towards the realisation of goals over and above the self-interest of the actor involved. However, there is a narrower philosophical use that involves a specific approach to moral questions. This is the so called 'virtue ethics' that is typically presented as a contrast to consequentialist and deontological approaches to moral problems. For those interested in increasing the other-regarding element of institutional design, and in particular law, virtue ethics potentially provides a rich resource.

Until recently, most moral theorists in the modern Western tradition have been either consequentialists or deontologists. According to the consequentialist, the rightness or wrongness of an action is determined by its consequence. Our aim in acting should be to maximise the good (whatever that turns out to be). Deontologists, on the other hand, are concerned with duties: we have obligations to perform certain actions in themselves or refrain from others, regardless of the consequences. Although very different, these theories share a focus on the formulation of moral rules or principles of right action.

Virtue ethicists, by way of contrast, focus on moral character and moral notions such as honesty, courage and patience. Here, the fundamental question an agent must ask themselves in deliberating morally is not so much 'Is my action in accordance with the favoured rule of morality?' but 'What sort of person should I be and what sort of person should I become?' We act morally as a consequence of our characters, rather than through some process of rationalisation. Following Aristotle, virtue ethicists see virtue as a 'habit of action': something that arises out of our natural dispositions; hence, the aim is to nurture a good character in oneself rather than developing the cognitive capacity to determine which course of action best adheres to an abstract principle. To possess a virtue is to be a certain sort of person.

To illustrate the difference, consider the various justifications these three competing theories might provide as to why one should help a person in distress. For the consequentialist, assisting the needy is a way of increasing the sum total of happiness in the world. For the deontologist, doing so is right because of various obligations we have to other morally significant beings (often understood as part of an obligation to treat each other as ends rather than mere means). For the virtue theorist, we do so because it is in our nature to help others. It is a consequence of character and thus the point is to develop the right kind of character.

Some critics have criticised virtue ethics for what they perceive to be its inability to be *action-guiding*. The objection runs as follows. According to the virtue ethicist, we should act virtuously, by which the virtue ethicists means doing what a virtuous person would do (one might ask oneself, 'what would Socrates do?', 'What would the Buddha do?', and so on). This is said to be circular because we have no independent criteria of what it is that marks out the virtuous person. However, as Rosalind Hurthouse (1991) notes, this objection fails to take into account the *action guidance* that is provided by specific virtue and vice terms such as honesty and charity. We should do what is honest/charitable and refrain from what is dishonest/uncharitable. Her claim is that through a focus on a range of virtue and vices, the virtue ethicist is able to provide action guidance.[8]

The relevance of virtue ethics to questions of what might be expected of natural resource users and how natural resource management might be improved should (hopefully) be clear. The idea would be that we encourage the development of specific virtues relevant to natural resource management, such as ideals of stewardship. Laws should be developed that make reference to such virtues and that hold resource users responsible for the development and exercise of environmental and social virtues. In the case of water users, a set of riparian virtues might be re-introduced in law. These would include such virtues as fair and judicious usage, honesty and a concern with stewardship of the water systems in question.

Applying virtue ethics to our legal frameworks

Virtue ethics offers an alternate approach that involves establishing ethical boundaries of responsibility by building on the utility of riparian rights and overcoming the limits placed on legal accountability to ensure ordinary and reasonable use. The purpose of such an approach is to offer a model of responsibility based on important elements of our ordinary practical reasoning. In essence, this suggests enabling responsibility through recognition of virtuous character traits (Hursthouse 2005).

Historical rights and obligation models of water management arguably have more characteristics, which reflect concepts of inter-relationship, awareness and responsibility for one's impacts and precaution, which fit with what modern ecologic science suggests is essential for sustainability. The commercialisation evident in water law through separation of land and water property rights, along with the introduction of tradeable water rights, has pushed water management further away from the conceptual model of sustainability. Under the market regime, legal rights in land and water have been separated, but environmentally, such links remain (Fisher 2004).

We now have an administrative system for water (and natural resources generally) that calls for obligations of virtue to improve stewardship. This incorporates notions of improved environmental protection and achieving ecologically sustainable development. The problem is that such expectations sit within a legal framework that protects the freedom of property holders to act with minimal accountability. This is a tension that is likely to lead to confusion about the practical meaning of responsibilities: a tension often hidden by terms such as 'stewardship'.

The concept of 'stewardship' – the guardian of place, holding a position of responsibility (Pearsall and Trumble 2001) – has become important in modern conceptions of natural resource use. It has been adopted in policies that advocate farmer responsibility for sustainable natural resource management (Barnes 2009; Carr 2002; Curry 2002). In relation to farming, the core duties of stewards are conservation to keep resources for posterity and protection to save resources from harm (Barnes 2009). Such responsibility is said by critics to be lacking in modern agricultural production systems (Baldock *et al.* 1996). Instead, industrial agriculture has been blamed for causing environmental decay (Beale and Fray 1990; Cocklin 2005; Curry 2002; Fullerton 2001; Roberts 1995). What is being increasingly advocated is good environmental practices in which farmers do not deplete resources as part of their obligation to future generations (Royal Commission on Environmental Pollution 1996). 'Stewardship' provides a conception of prudent or right behaviour to limit or reverse environmental harm (Lee 2005). Prudence is about ends: how to make important choices using a mixture of foresight, morals and self understanding; in effect a demonstration of virtue (Jacob 1995).

The discourse of stewardship seeks constraints on exploitation, in the public interest. This concept acts to limit the exploitative freedom implicit in property rights. It helps form norms of conservation practice, and protects legitimacy and social trust in return for environmentally benign farming practice.[9] It is a virtuous conception of performance supportive of social licence and characterised by higher conservation standards. Figure 3.1 illustrates the distinction between minimum accountability and virtue.

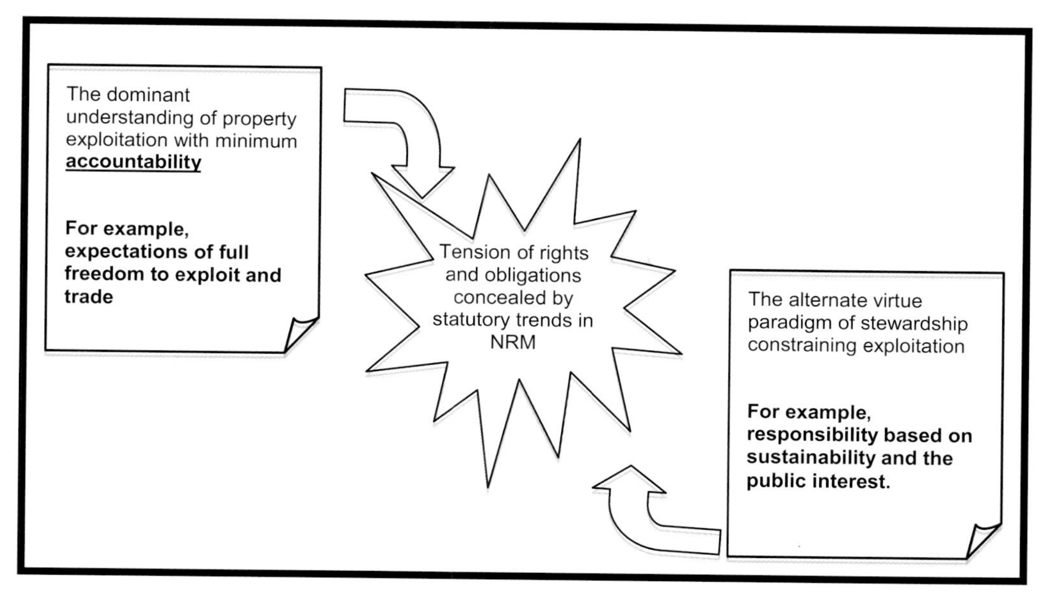

Figure 3.1: Tension between competing expectations for natural resource management

At a practical level, a failure to resolve the tension leads to a range of interpretations about the meaning of 'stewardship'. These represent competing expectations about the legally enforceable boundaries of responsibility for natural resource management by farmers. Such tensions are brought into focus when a practical meaning needs to be defined (Cocklin *et al.* 2007; Crosthwaite 2001). The difficulties have been illustrated elsewhere by reference to a statutory precautionary principle in decision making and the statutory duty of care applied to farming (Shepheard and Martin 2009).

Law, virtue and responsibility in natural resource management

Conceptualising resource access through the use of (property) rights within a framework of necessary moral values runs counter to conceptions which involve the severing of property rights from environmental and social values. Property rights are conceptualised as utilitarian interests, curbed only by rules and policies of the State where interests clash (Coyle and Morrow 2004). The impact of market approaches has been to reinforce the separation of property rights in land from those in water, developing separate commodities or interests, despite the obvious physical and ecological connections between the two (Fisher 2004). Our purpose in introducing moral values and expecting virtuous behaviour on the part of users is to suggest a theory of property, based on responsibility as much as right. It is this point that we wish to take up as a focus for exploring the role of virtue in our western liberal legal framework. This view reinforces the idea that moral virtues are a necessary part of a just and efficient legal order (Koller 2007) and that law provides a means to clarify and deepen the moral meaning of human situations (Novak 2009).

For farming, a primary role for law is to define the responsibilities of access to, and use of, natural resources such as water. These can be structured around accountability as a minimum required level of behaviour, or virtue as a desired standard of ethical performance. This is a distinction pertinent to global expectations about sustainability and universal responsibility. In the discourse of international conventions, concepts of stewardship abound. However, in the implementation at a state level, the discourse property as the mechanism to meet these aspirations is equally strong. The reconciliation between these discourses is, as this chapter has suggested, problematic.

Rights exist as a legally supported relationship between people governing control and access to resources. Within this relationship, there is potential for virtue to shape relationships that reflect moral expectations to the broader community, and to immediately affected neighbours. Virtue can thus be of great practical use in meeting social responsibility suggested by a social licence.

Social responsibility debates are important to irrigation businesses (Gunningham 2004). Water reform has changed how the water resource is shared, making the trade-offs between the environment, urban areas and farming more apparent. Water enterprises are continuously engaged in a negotiation with society. Justification of the social licence is continuously required to succeed in these negotiations, suggesting that in the longer term the exercise of private property interests is materially constrained by accountability to the community. These tensions suggest that in practice in society, regardless of the apparent certainty of property rights, users of resources are subject to social expectations of some level of virtue. Further, there are indications that claims of virtue are politically important in supporting claims for the protection, exercise or extension of legal property rights.

Boundaries of responsibility are formally specified by law, but also informally subject to social licence (we can regard social licence as involving an expectation of virtue). The reality and significance of this dynamic can be observed in the processes through which water is actually allocated. A water entitlement as a form of property involves two property rights. One is the (tradeable) licence to extract some percentage of the available water, with availability being administratively determined. The second is the use right, once that water is available. These rights are continuously adjusted through mixed political, legal and administrative processes. These include the negotiation of water sharing plans and decisions about annual allocations, the development of laws to determine the priority of water access, and public investment in water infrastructure. These processes determine the conditions for trading, use and the availability of water. Many changes in access to water occur that call into question the apparent security (or property right) that a tradeable entitlement to water suggests (Shepheard and Martin 2008).

Social licence, virtue parsimony and riparian rights: some concluding remarks

Recent debates in the social licence literature are reflective of a growing sense of disquiet that something is lacking in our legal frameworks with respect to what it is that society expects from resource users. Such disquiet is at odds with models of law that privilege property exploitation and minimal accountability (the latter we might think of as virtue parsimony embodied in law). We can see the influence of these models in recent developments in water law, where there has been an inexorable move away from riparian rights, that involve an expectation of moral concerns with resources used, to modern conceptions of water rights involving exclusive use.

Debates over social licence in the law (and more specifically in water law) mirror broader tensions between community expectations of virtue of those entrusted with public resources and popular economic models of institutional design, which have no such expectations. In this chapter, we have explored some ways in which natural resources law might incorporate expectations of virtuous behaviour on the part of resources managers and, in this way, bring the law closer to the ideals of social licence that are implicit within a great deal of public discourse surrounding sustainability and stewardship. The tension between the apparent certainties of formal property rights, the countervailing legal obligations for responsible resource management, and the increasing social expectations of stewardship is a major challenge for farmers and for policy makers. This chapter has attempted to highlight the role of concepts of virtue that is hidden within the contests that are triggered by these tensions. We have suggested that a more open recognition of responsibilities generated by social expectations of virtue would help to recognise the realities of our complex society, and the complex legal relationships that are the result. We trust that this chapter will contribute to bringing law and policy into a closer alignment with the realities of changing community expectations.

Endnotes

1 Note that it is the virtue of the citizens which is our focus here. A different, but closely related question concerns whether institutions *aim* to produce virtue in their citizenry.
2 See Robertson D (1956) What do economists economise? In: *Economic Commentaries.* Staples Press, London.
3 Smith is merely talking about self-interest, rather than vicious malevolent behaviour.
4 This refers to common law riparian rights, which should be distinguished from riparian rights under the *Water Act* 1912 (NSW).
5 See Brennan and Hamlin (1995), p.35–56.
6 See, for instance, Carter and Maddock (1987).
7 See, for instance, Bowles S (2008) Policies designed for self-interested citizens may undermine 'the moral sentiments': evidence from experiments. *Science* **320** (5883), 1605–1609.
8 As Hursthouse (2005) notes: 'Much invaluable actions guidance comes from avoiding courses of action that would be irresponsible, feckless, lazy, inconsiderate, uncooperative, harsh, intolerant, selfish, mercenary, indiscreet, tactless, arrogant, unsympathetic, cold, incautious, unenterprising, pusillanimous, feeble, presumptuous, rude, hypocritical, self-indulgent, materialistic, grasping, short-sighted, vindictive, calculating, ungrateful, grudging, brutal, profligate, disloyal and so on.'
9 These are boundary-setting dimensions that are relevant to defining natural resource access and use rights for farmers.

References and further reading

Arrow K (1971) Political and economic evaluation of social effects and externalities. In: *Frontiers of Quantitative Economics.* (Ed. M Inriligator) pp. 3–25. North-Holland Press, Amsterdam.

Baldock D, Bishop K and Mitchell K (1996) 'Growing greener: sustainable agriculture in the UK'. World Wide Fund for Nature, Godalming, UK.

Barnes R (2009) *Property Rights and Natural Resources.* HART Publishing, Oxford.

Beale B and Fray P (1990) *The Vanishing Continent. Australia's Degraded Environment.* Hodder & Stoughton, Sydney.

Bowring J (Ed.) (1843) *The Collected Works of Jeremy Bentham* Vol 8. William Tait Publishers, Edinburgh.

Brandt RB (1976) The psychology of benevolence and its implications for philosophy. *Journal of Philosophy* **73**, 429–453.

Brennan G and Hamlin A (1995) Economising on virtue. *Constitutional Political Economy* **6**, 35–56.

Buchanan J (1975) *The Limits of Liberty: Between Anarchy and the Leviathan.* University of Chicago Press, Chicago.

Carr A (2002) *Grass Roots and Green Tape. Principles and Practices of Environmental Stewardship.* Federation Press, Sydney.

Carter M and Maddock R (1987) Inflation: the invisible foot of macroeconomics. *The Economic Record* **63**, 120–128.

Cocklin C (2005) Natural capital and the sustainability of rural communities. In: *Sustainability and Change in Rural Australia.* (Eds C Cocklin and J Dibden) pp. 171–191. University of NSW Press, Sydney.

Cocklin C, Mautner N and Dibden J (2007) Public policy, private landholders: perspectives on policy mechanisms for sustainable land management. *Journal of Environmental Management* **85**(4), 986–998.

Coyle S and Morrow K (2004) *Philosophical Foundations of Environmental Law, Property, Rights and Nature.* HART Publishing, Oxford.

Crosthwaite J (2001) 'Farmer land stewardship: a pillar to reinforce natural resource management?'. *Connections Farm, Food, Resource Issues* Vol 1 December 2001.

Curry D (2002) 'Farming & food. A sustainable future'. The Policy Commission on Farming and Food, London.

Fisher DE (2004) Rights of property in water: confusion or clarity. *Environment and Planning Law Journal* **21**, 200–226.

Frey B (1997) A constitution for knaves crowds out civic virtues. *The Economic Journal* **107**, 1043–1053.

Fullerton T (2001) *Watershed. Deciding Our Water Future. Juggling the Interests of Farmers, Politicians, Big Business, Ordinary People and Nature.* ABC Books, Sydney.

Goodin RE (1989) Do motives matter? *Canadian Journal of Philosophy* **19**, 405–420.

Goodin RE and Philip P (Eds) (1993) *A Companion to Contemporary Political Philosophy.* Blackwell, Oxford.

Gunningham N (2004) Cotton, health and environment: a case study of self-regulation. *The Australasian Journal of Natural Resources Law and Policy* **9**(2), 189–228.

Hahn F (1991) Benevolence. In: *Thoughtful Economic Man.* (Ed. J Gay Tulip Meeks) pp.7–11. Cambridge University Press, Cambridge.

Hamilton A, Madison J and Jay J (1788) *The Federalist.* (Ed. JE Cooke) Wesleyan University Press, Connecticut, 1961.

Hirschman A (1984) Against parsimony: three ways of complicating some categories of economic discourse. *American Economic Review* **7**(2), 89–96.

Howarth W (1992) *Wisdom's Law of Watercourses.* 5th edn. Shaw and Sons Limited, Crayford, Kent.

Hume D (1741) Of the independency of parliament. In: *Hume Political Essays.* (Ed. K Haarkonssen) pp. 24–27. Cambridge University Press, Cambridge (re-published 1994).

Hurthouse R (1991) Virtue theory and abortion. *Philosophy and Public Affairs* 20 (3), 223–246.

Hursthouse R (2005) Are virtues the proper starting point for morality? In: *Contemporary Debates in Moral Theory.* (Ed. James Dreier) pp. 99–112. Wiley-Blackwell, Oxford.

Jacob BE (1995) Ancient rhetoric, modern legal thought, and politics: a review essay on the translation of Viehweg's 'Topics and Law'. *Northwestern University Law Review* **89** (summer).

Johnson P (1988) *Politics, Innocence and the Limits of Goodness.* Routledge, London.

Koller P (2007) Law, morality, and virtue. In: *Working Virtue. Virtue Ethics and Contemporary Moral Problems.* (Eds RL Walker and PJ Ivanhoe) pp. 191–206. Clarendon Press, Oxford.

Lee M (2005) *EU Environmental Law: Challenges, Change and Decision-Making: Modern Studies in European Law 6.* HART Publishing, Oxford.

Novak D (2009) Natural law, human dignity and the protection of human property. In: *Profit, Prudence and Virtue. Essays in Ethics, Business and Management.* (Eds S Gregg and J Stoner) pp. 42–56. Imprint Academic, Exeter.

Pearsall J and Trumble B (Eds) (2001) *Oxford English Reference Dictionary.* 2nd rev. edn. Oxford University Press, Oxford.

Roberts B (1995) *The Quest for Sustainable Agriculture and Land Use.* University of NSW Press, Sydney.

Robbins L (1935) *An Essay on the Nature and Significance of Economic Science.* 2nd edn. Macmillan, London.

Royal Commission on Environmental Pollution (1996) 'Sustainable use of soil'. HMSO, London.

Sen A (1991) Beneconfusion. In: *Thoughtful Economic Man.* (Ed. J Gay Tulip Meeks) pp. 12–16. Cambridge University Press, Cambridge.

Shepheard M and Martin P (2008) Social licence to irrigate: the boundary problem. *Social Alternatives* **27**(3), 32–39.

Shepheard M and Martin P (2009) Multiple meanings and practical problems: the duty of care and stewardship in agriculture. Macquarie Journal of International and Comparative Environmental Law **6**, 191–215.

Smith A (1776) *The Wealth of Nations.* Random House Inc, New York (republished 2000).

EXPERIENCE OF FARMERS

4

Organic poetic licence: consumer moral norms driving farming systems

Andrew Monk

This chapter will look at the organic food and farming industry internationally, with a focus on Australia. Organic farming has developed and matured from a movement merging farmer and consumer interests, to become an established niche market offering in mainstream supermarkets. 'Organic' has achieved strong consumer support, creating a strong social licence that is directly reflected in consumer purchasing preferences, to roll out its philosophy and practice. This is done according to clearly articulated standards and independent auditing and verification, and has in large measure created a growing consumer following, albeit representing a niche of the overall market place for food and farming. Further, both consumers and producers have become deeply committed to the principles of organics, with this aesthetic or emotional depth of commitment being reflected in the idea of a *poetic licence*. The organic sector as a case study demonstrates that the combination of strong self-governance using objective and independently audited standards, setting these standards at a high level and communication to the consumer, are all key elements in defending and expanding the reputation and therefore the social licence of farmers.

What lessons can agriculture generally learn from this example, particularly about the extent to which the market can 'pull through' improved farming sustainability practices and support industry codes, providing an economic rationale for improved farming practice? The answer I suggest lies in ensuring that any move to generate social licence to farm, let alone poetic licence, to enlist consumers to be passionate about one's cause must revolve around a package of goods that are delivered in a convincing, meaningful and transparent manner. This is no mean task and the organic industry and movement has been doing this for over three decades in a formal market sense.

The success of the organics case study in Australia relies on two fundamental pillars. One is broad stakeholder involvement (the *process*) and the other is the package of *goods* that are delivered, and/or perceived to be delivered, by the process (the *content* or *standard*). Although one can exist without the other, they can fast become orphans without the mutually reinforcing roles that both pillars have to play in delivering both confidence and trust in key stakeholders, from farmers through to consumers.

The success of the organics approach can be best understood within the context of the *certification revolution* articulated by Michael Conroy (2007). The independence of production standards-setting processes with broad and meaningful stakeholder input, combined with the professional independence of auditing and certification is key to success in supporting the future licence to farm, by ensuring well-justified consumer confidence in the producer (Conroy 2007).

Irrespective of one's views and the scientific and philosophical (and sometimes very emotional) debates about organic farming and food versus other forms of production, the traction that this movement has achieved is worthy of study in its own right. It suggests a successful pathway for meaningful dialogue with stakeholders, and the establishment of responsive markets as a driver for change and/or the protection of desirable practices in the field.

This chapter outlines the successes of these processes and the pitfalls that face any such radical agenda that attempts to link rewards in the market place with improved producer behaviour in the field. The audacious agenda of the organic movement and its certification industry has many resistance points and challenges. However, it continues to garner allies, including the largest retailers, and now producers and processors, in the world. Its successes, as well as failures and shortcomings, are useful beacons for others pursuing similar aims in other fields.

The risk of fragmentation: a good endorsement program is not the sum of parts

The challenges posed by the schizophrenic behaviour of the modern consumer – who is bedazzled by a range of issues and offers, which are often reduced to simplistic and sometimes jingoistic 'solutions' to their concerns – are many fold. The modern agri-food system has also become schizophrenic as it attempts to react to, and sometimes co-opt, an array of messages from and to consumers about concerns ranging from animal welfare, environmental protection, water efficiency and contaminants in products to, more recently, the ubiquitous question of a carbon footprint.

Media and government agencies are quick to isolate issues for attention. This in turn tends to generate a whole industry reaction to singular fads. In this vicious cycle are generated many well-meaning, but often confused, attempts at building legitimacy, and therefore in turn to recreate the licence for producers to (re)win favour with their constituents. That social licence from the consumer is directly reflected in their purchasing preferences, and therefore in the income of producers. Achieving it is also reflected in the way that products are produced, which in turn affects the operations and economics of farming.

Markets have reactively struggled to deliver branded solutions for consumers that connect a claim or product endorsement with perceptions of the consumer's views of the social licence to farm. The term *green wash* may more aptly be termed *green saturation,* as supermarket shelves burst with all manner of environmental and social benefit claims, confusing the consumer and leading to ever more scepticism about what is true in all those claims. Compounding this will be the ongoing inevitable confusion and complexity of determining sustainability merits based on water, soil and carbon efficiency, and related sustainability merits of different products and processes. This in turn may make it more difficult, in the future, to use market-labelling mechanisms to achieve consumer support to leverage producer behavioural change on the basis of demonstration of the farmer's satisfaction of the consumers' expectations around the social licence to farm.

The organic paradigm

Enter the notion of organic that somehow has captured a small, but growing, segment of the consumer population and has forced larger supermarkets to take note and stock up. Organic as a concept, enshrined internationally in standards, regulation and certification, has delivered not just legal licence but captured the high moral ground, creating a *moral licence to farm*. This has allowed some producers to claim a status that commands a premium in the market place.

At this level, it has been then a relatively simple step to creating one's own rules (poetic licence) intended to protect that moral licence and the associated market opportunity.

The organic farming movement can trace its roots from the middle of last century, with the reaction to the rising use of synthetic pesticides and increasing corporate or agribusiness trends in farming. The other major influence was a sense that consumers were becoming disenfranchised from being able to make active and informed choices about their food, where it came from, who produced it and under what methods. From the 1970s, there were more formal approaches taken by farmers and consumers, who came together to establish standards by which their farming and food production systems would be governed: banning synthetic pesticides, shunning productivity pressures over sustainability and ecological priorities, and re-orienting the status of farm animals from production 'units' back into integrated parts of the whole 'organism' of the farm.

Whether it is stories of herbicide residues in watertables, pesticide residues in meat products banned at an importing country port, or the short media snippets on animals in pens or cages, such issues resonate with increasingly urbanised (and therefore removed from the action) consumers who are wary about what is going on in their food chain. Organic is therefore positioned as a 'patron saint', taking a stance of ameliorating or preventing such events for consumers willing to buy its message. It is seen as reducing the risks of consumers, who do not understand well what those risks might be, but who are concerned to avoid harm to themselves and their families, and harm to the environment or to less-advantaged people.

Organic farming now has well-entrenched international standards, with both government and industry or movement-driven regulatory and verification arrangements derived from quite radical beginnings in the 1970s. The organic farming movement became a successful industry and market niche through the 1990s, in particular via the uptake of organic food products in the major supermarkets of Europe and the US. Standards around the world (e.g. EU 824/2007 and USDA NOP[1]; Codex; IFOAM) have locked it into a clearly articulated and recognisable framework and set of core commitments.

The processes of standards setting, independent assessment and verification, and market monitoring remain to this day both 'organic' and true to the core of the original tenets of the movement of the 1970s. It is arguably this discipline – of remaining focused on processes of standards setting and maintaining broad stakeholder input into these processes (rather than being beholden to singular commercial interests or interest groups) – that maintains organic as a distinct and sought-after food category in the markets of the developed economies. Consumers, it seems, do trust this stewardship integrity and this is at the heart of their willingness to select certified organic over other products.

Like the food industry generally, 'organic' as a category is not without its contradictions and ethical challenges. The upside has been that it has seen the conversion of a percentage of farmers in developed economies convert their farmlands to organic production practices, and in turn given access to consumers wanting produced pesticide-free or, more recently, genetically modified organism (GMO)-free food products. In an environment where consumers are arguably getting more confused rather than better informed about food labels, food origins and food ethics, coupled with more farmers continuing to leave the land because of lowered returns and competition from anonymous producers in other countries, the growth of the organic market and its networks of farmers to consumers stands out as a case of successful social change engineering.

The organic movement's challenge remains the balancing of the competing agendas and interests of government, big business and commercial retailers who seek to maximise their financial returns and market share, with that of consumers and producers, while remaining true to the ideals embodied in the standards for organic production. Organic processes of standards setting themselves at their best remain 'organic' and ever-changing rather than ossified. The organic industry is a case study of producer–consumer-driven food production regulation in action. At the centre of this success are resilient and sustainable industry associations and social structures. These provide the glue to manage the conflicts, as well as opportunities for further social or practice change in the food and farming community. These observations suggest that the dynamics of the social processes – coupled with a commitment to particular norms and beliefs that are reinforced through these interactions – are at the heart of the success of the organics movement in carving out a position of moral licence, which is in turn reflected in an unique market position.

Organic standards and sustaining representation of the stakeholders

The fundamental strength of the organic movement has not been its stance against synthetic pesticides or less-than-ideal animal welfare standards, but the manner in which it goes about reviewing its stances on food and farm technologies, and establishing and maintaining standards and systems to deal with these. The case of GMOs and their progressive integration into the food and farming system through the 1990s is a classic example of technologies being implemented arguably against the wish of a significant proportion of consumers. These consumers, without a trusted certification brand, are otherwise unable to make discerning choices at the point of retail about the use of such technologies in making their own food purchases.

The organic industry has been able to establish a 'choice' for consumers amidst this sea of confusing, and sometimes opaque, labelling requirements (or lack thereof). This in turn creates a channel to market for farmers and 'value adders' wishing to supply to that demand.

Well-established standard-setting processes, via consultation with industry members, consumer groups, food technologists and others through the 1990s, determined that GMOs did not have a place in organic production principles and therefore in organic farming. This stance has been reinforced again and again in a variety of international settings. The most extreme example of social involvement in organic standards setting was the 1990s processes in the US. This saw a large public response through submissions to a draft USDA organic standard, which suggested, among other things, the inclusion of GMOs in the organic standard. Over 250,000 submissions were received on this draft, with a resounding 'no' vote being registered by most. The consumer's aversion to the risks of GMOs is well established, even in the face of industry claims that the consumer's perceptions are ill-founded. Because the organic brand serves the purpose of signalling to consumers that things they do not want in the products they buy are indeed not present, regardless of any science debate, this suggests the importance of organic standards providing the consumer with confidence that their preferences are being met by producers.

The level of social engagement within the organic community on such issues, and the regular commentary that the organic movement has on food and farming practices more broadly, continues to provide the social 'gel' that ensures a continued focus on the issues. It is through the organic certification bodies that consumer voices are also heard in formal government processes, providing the resources (both financial and social) such that submissions are made, the public is informed via the media and other social forums, and governments are pressured to deliver on policies that align with the interests of the consumer.

This success in building trust has translated into premium prices for organic foods being maintained at the retail level. This allows farmers to continue to maintain what are very exacting agricultural and food standards that meet the expectations of the consumers who demand them. An industry-driven and maintained system of auditing or inspection and certification to the standards ensures that commercial pressures and self interest do not outweigh and take over the principles upon which the standards are based in the first place.

On the one hand, the organic industry's success in market share growth and ongoing uptake by consumers and farmers alike (albeit still at niche levels of low single digit percentages of total market)[2] continues to enable it to have influence on food standards and some production methods. On the other hand, such success and the extension of the social network of the organic industry risks it becoming beholden to the masters and market pressures that gave birth to the movement in the first place.

The world's largest food companies now own organic processing facilities and organic brands (e.g. Danone, General Mills and Proctor and Gamble, as well as more locally in Australia, with Heinz, Sanitarium, National Foods and the major retailers Woolworths and Coles). In some countries there has been a significant takeover of standard setting arrangements by government (the most significant being the US via the US Department of Agriculture). This has benefits, but also generates tensions because commercial interests at times clash with organic ideals of production. These various entities are competing for a share of the consumer's trust, as precursors to a share in the consumer's spending. Publications, such as *Organic Inc* have chronicled how these tensions and pressures continue to play out in the US, with implications for other countries including Australia in the decades ahead.

Sitting in-between market (retail and producer) pressures on the one hand and increasing government involvement are industry associations: the buffering agents who facilitate ongoing negotiation between market needs and the organic movement principles and norms. Not all associations are born, or are maintained, equal or organic. Some become co-opted by the very pressures they are set up to manage. Others fail to generate a significant legacy or impact owing to unsustainable structures in the form of (or lack of) social networks, corporate governance or financial arrangements. Industry associations – owned by industry members and structured to ensure equal voting of members, while also having a structure that limits overall commercial pressure from a single individual, group or sector – are arguably one of the most resilient means of ensuring that these competing pressures do not negate or destroy the ideals upon which such a movement as organic is predicated.

The case of the Biological Farmers of Australia Ltd (BFA), now in its third decade of operation, is a case in point. BFA is a large and successful organic industry representative, advocacy and services association. It is the institutional home to over half of the members of the Australian organic industry through their voluntary membership. It has subsidiary operations that conduct independent auditing and certification to standards set by its members and stakeholders. The BFA has a structure that caps financial contributions to a level such that any one commercial interest or individual cannot contribute more than 1% of overall turnover of the group. Most importantly, it has standard corporate structures that enable voting of a Board of Directors on a rotating biennial basis, as well as advisory groups and stakeholder groups that ensure that broad consultation is conducted. This is to enable transparent and effective decision making, whether of a technical or policy nature. It invests significantly in regional and metropolitan *roadshows* and other networking events that ensure that members of industry have the opportunity for input and involvement throughout the year on a variety of initiatives and activities.

The recent entry of Standards Australia into organic standards setting, with the publication of the *AS 6000–2009 Organic and Biodynamic Products Standard*, has further underscored the industry's independence and broad stakeholder input into standards setting. By no coincidence, this standard reflects the ideals and specifications articulated by industry associations such as the BFA and the consumer and producer communities that they represent. The recipe for success for the organic industry in Australia, albeit an unfolding and never-ending process, relies on balancing competing agendas and stakeholder interests with a pragmatism for using existing social and financial structures (i.e. retail markets and government rules) to influence farmer and retailer behaviour alike.

The very success of the organic movement, and the litmus test of its future performance, will be how well it maintains these social networks for consultation with stakeholders, the balancing of competing and sometimes conflicting interests, the tempering of overt commercial and government interests and agendas, and its pragmatism in remaining relevant to and for the general public. All of these cannot all be achieved simultaneously all the time. Like democracy itself, the organic movement's strength and effectiveness can never be taken for granted. It requires continual renewal and engagement by affected stakeholders.

The defining point about 'organic', as a movement, as a standard and as a formal certification mark, is that its bedrock is that 'we make a stand and we try'. There is significant transparency to its make-up and a broad array of stakeholders involved in the standards-setting process. It is not beholden to one sectoral interest, and remains to this day with an uneasy but nonetheless ongoing search for equilibrium between big and small business interests.

Ultimately, a small team of individuals and groups that are best placed and able to continue to engage with the wider stakeholder population will be required to shape this consumer-focused moral licence movement. It is simply not feasible to operate without efficient institutional structures to handle the complex acts of consultation, management and governance that are involved. The role of well-structured, pragmatic and well-resourced associations and social groups are key in this process, and their importance cannot be over estimated. Investing significantly in getting these structures right early on, and maintaining them with the ownership, *buy in* and support of the pivotal stakeholders, which must include consumer interests, is key to lasting success. It is also the key to achievement of the ongoing social change that is reflected in the ideals of the movement that is directing and inspiring this.

The complexity of balancing interests and conducting a multi-facetted dialogue in order to maintain a credible standard is similar to how *Forestry Stewardship Council* stakeholders are situated. Strong stakeholders pushing for stringent standards are essential, or there is no basis for arguing for legitimacy (and therefore social licence) claims. However, this has to be managed along with ensuring for producers that the standards are feasible to implement in a manner sufficient to achieve sufficient uptake. There has to be a sufficient scale of both demand and supply at the level of standard that is set, to ensure that the certification scheme is viable and attractive.

The recipe for success in printing licences

The key ingredients for success for the organic movement and industry have been first supportive and well-resourced industry-owned institutions, embodied in non government organisations (NGOs) such as the International Federation of Organic Agriculture Movements (IFOAM) and within Australian groups such as the BFA. Second, the relationship networks, such as non-government organisations, have been able to construct with the world of supply chains and retailers, coupled with the presence of well-resourced independent auditing and certification programs that verify through the chain the claims and endorsements that are being made.

These are the ingredients that Michael Conroy (2007) in his study of the globalisation of certification programs and product endorsement branding claims notes have been the basis for the success of such programs, and the reason they may survive or fail in the future. Conroy notes three ingredients needed to maintain an equilibrium so as to satisfy the ongoing goal of legitimacy and impact (and therefore the chance for maintaining the benefits of a strong social licence): NGOs that can identify and diplomatically exploit corporate 'soft spots'; corporations, farming groups and individuals who recognise their vulnerability and seek ethical solutions; and evolving and effective independent certification programs that continue to be exactly that.

The poetic licence that the organic movement has enjoyed, and in turn the industry that has evolved around it (including the standards and certification industry), is not a guarantee of immunity from future prosecution or persecution from a critical audience of consumers. This movement has weathered a number of storms, taken some fights front on and faces challenges from many sides (including inside disagreements on philosophy or policy, as is the case with any passionate environmental movement base).

The potent mixture, however, of market support coupled with consumer and environmental (i.e. non producer/business interest) stakeholder involvement is a sure-fire means of giving it the best chance of success into the future. This requires a rich dialogue, and a willingness to truly listen to and respect what the consumer says through that dialogue. It is these elements that we all should take note of in attempting to establish both legal and *moral licence to operate farming activities* in any setting, in any sector, in any country of the world.

Endnotes

1 EU: Regulation EU 834/2007; USDA NOP; US Department of Agriculture National Organic Program; JAS: Japan Agricultural Standard for Organics.

2 See Mitchell *et al.* (2010) 'Australian organic market report'; Organic Trade Association.

References and further reading

Biological Farmers of Australia Ltd (2010) *Australian Certified Organic Standard.* BFA Ltd, Brisbane.

Conroy M (2007) *Branded! How the Certification Revolution Is Transforming Global Corporations.* New Society Publishers, Canada.

Mitchell A, Kristiansen P, Bez N and Monk A (2010) 'Australian organic market report 2010'. BFA Ltd, Brisbane.

5

Triple bottom line reporting in the irrigation sector

Evan Christen, Mark Shepheard, Wayne Meyer and Christopher Stone

Most irrigation in Australia was instigated with regional socio-economic development as a primary goal. The irrigation schemes were developed with grand visions of settling the interior and providing farming opportunities to soldiers returning from world wars. Thus it could be said that the master planners in government had a fledgling version of the triple bottom line in their sights. These visions have partly been realised with evidence from the Murrumbidgee and Murray basins of significant inland populations associated with the irrigated districts and an annual revenue of $3.1 billion associated with an investment into irrigation infrastructure valued at about $10 billion (Meyer 2005).

This economic development in the Murrumbidgee and Murray basins brought about by this irrigation uses about 8600 Gigalitres (GL) of water, which is about 52% of the total annual runoff into the system. This level of water extraction has led to ecosystem problems associated with highly controlled river flows and a disconnection of the river from the floodplain ecosystems. Public and government concern has resulted in actions such as the National Water Initiative (NWI) and Murray Darling Basin Plan to recover and return water for targeted environmental flows.

Irrigation development induces considerable environmental change, but the expectation has been in the past that the economic and social benefits would be greater than the environmental costs. However, public attitudes have changed over time, from enthusiasm for development and exploitation to greater concern regarding environmental issues and sustainability. Recently, the irrigation industry has found it difficult to communicate to the wider populace and government the benefits of irrigation to their regions and to explain the activities and investment undertaken to tackle the environmental sustainability concerns. To rectify this, irrigation water supply businesses are investigating using a reporting structure that includes financial, environmental, and social and cultural elements. This triple bottom line, holistic approach is intended to provide a more balanced view of water use, with socio-economic benefits and environmental consequences demonstrated. It is anticipated that this approach will lead to a more transparent and informed debate on the sustainable use of resources between all parties.

This chapter will provide an overview the history of irrigation in the Murray and Murrumbidgee basins, their current environmental and socio-economic conditions and the context for sustainability performance reporting by irrigation water suppliers. Two case studies of irrigation company performance reporting will be presented. The concept of sustainability introduces expectations of a *social licence* as practical concerns for irrigation water supply businesses. These concerns about demonstrating responsible performance go beyond the legal reporting requirements for annual financial and environmental compliance reporting.

Case study: the development and effects of irrigation

The development of irrigated agriculture is an illustration of a combination of community attitudes to natural resources, individual entrepreneurship and effort, and government prioritisation and investment. The social licence to access funds and resources is pivotal to the 'opening up' of new irrigation areas. Making use of the available water resources triggered the imagination of many early settlers. The earliest irrigation developments in Australia were undertaken in the coastal settlements of New South Wales (around Sydney) and in Tasmania (Blackburn 2004). In the mid 1800s, pastoral development (primarily grazing of sheep) began on large flat inland tracts of south-central and south-western New South Wales and northern Victoria. Grazing sheep for wool was jeopardised by the frequency of drought, making access to waters of the inland rivers critical for the survival of these settlements.

Agitation for insurance against the effects of drought through irrigation was in part responsible for the formation of the Deakin Royal Commission of Victoria in 1884. Deakin visited the USA, India and other irrigation areas and strongly recommended the development of large irrigation schemes in Victoria. This resulted in the *Irrigation Act 1886* (Vic). This ensured all water rights rested with the government and, to a large extent, committed it to carry out major water storage and regulation works. Similar laws followed in other states (Hallows and Thompson 1995).

Early attempts at forming private Irrigation Trusts were largely unsuccessful, returns from water sales were inadequate and returns from irrigated production were generally poor. Irrigation schemes did proceed though largely with government backing and control and with a focus on support for the pastoral industry. At the instigation of the state governments of South Australia and Victoria, the Chaffey Brothers' developments at Renmark and Mildura proceeded with optimistic enthusiasm. Within a decade, however, both moved from grand vision to difficult implementation to effective failure caused by fundamental problems such as: the cost of lifting water from the river; seepage losses from unlined channels; poorly adapted crops; lack of land transport; and poor farming skills. But persistence prevailed and both areas are now productive irrigated regions, largely due to the determination of individuals to overcome problems and government willingness to inject capital into infrastructure, research and extension (Mack 2003).

These early developments set the framework for the irrigation industry in Australia. The large irrigation areas of south-east Australia were developed with public investment in dams and delivery infrastructure as a form of nation building. Subsequently, following the two world wars, soldiers were settled in irrigation schemes – deliberate social engineering to provide farming opportunities for returned servicemen. With few exceptions, most irrigation development has been motivated and supported by social and political ambitions.

The consequences of public enthusiasm and private enterprise

The period from around 2005 to 2010 has been marked by historically abnormal droughts, which have cut irrigated production dramatically from the long-term trends and averages. However, the total area irrigated in the Murray and Murrumbidgee basins is half of all the irrigated area in Australia. The regions in the basins use about half (8600 GL) of all the water used for irrigation in Australia (16 700 GL) to produce about one-third ($3.1 billion) of the total farm gate revenue. In addition to the extensive grain and fodder cropping, these regions produce one-third of all the fresh fruit and vegetables produced in Australia and two-thirds of all the wine, table and dried grapes (Meyer 2005).

The total population of the Murray and Murrumbidgee regions approaches 600 000 people, most of whom live in large towns. Comparison with adjacent rain-dependent districts shows that the addition of irrigation increases the level of economic activity, and the population that it supports, by three to five times. Without irrigation, these regions would have far fewer people, fewer large communities and fewer public services. The demographic profile of the irrigated regions shows populations that are more diverse in ethnic origin and with a greater proportion of younger people relative to the rain-dependent districts (Meyer 2005).

Irrigation development has changed the appearance of large tracts of land in the Murray and Murrumbidgee basins. The large infrastructure investment has brought productive agriculture and community growth to semi-arid inland Australia. On the land, native vegetation has been removed, wetlands drained or flooded, earth moved and drainage lines changed and soils cultivated. The extensive clearing and subsequent addition of large volumes of water have caused a fundamental change in groundwater distribution. In the rivers, flow patterns and volumes are very different from natural conditions, with return of drainage waters contributing salt, nutrients and chemicals.

There are clear signs that not all of these changes have delivered a net benefit. There are substantial areas where additions to groundwater have caused the unconfined aquifers to rise, forming a watertable mound. Salinisation of topsoils has significantly reduced the productive capacity of the land. Saline groundwater discharges have not only affected some low-lying areas in the irrigated regions but also contribute considerable salt loading to the rivers. These effects, combined with the very large change in the flow and seasonality of the Murray and Murrumbidgee Rivers, have caused large changes in river ecology and the connected floodplain and riverine ecosystems.

The combined effects of infrastructure development and water extraction for irrigation, return drainage, groundwater discharge and accompanying salt loads and reduced flow, especially of small and medium floods, have caused highly publicised changes to the rivers. Concerns of downstream water users, together with a heightened sense of environmental awareness, have raised the issues of water quantity, water quality and river health to state and national political levels. There is now a reasonably well-developed position, particularly among the large urban population of the major eastern state cities that irrigation practice needs to change to achieve a better balance between use of the water for production and use of water for environmental maintenance. This public shift in perceptions of irrigation has had fundamental effects upon the irrigation sector. It has translated through political process into new national water policies (the National Water Initiative) and laws (*The Water Act 2007* (Cwlth)). These developments have set ecological limits on water use, making farming use subordinate to ecological demands. They have also triggered billions of dollars of additional public and private investment in water use efficiency and the 'clawback' of large volumes of water from production. The link between social licence and economic viability has become very clear to any reflective observer of the dynamics of the irrigation industry.

It is against this background that irrigation communities and irrigation water supply companies are looking for approaches to assist in improving their performance and demonstrating transparently the total benefits and costs of irrigation. By doing this, a supportive debate and policy implementation can be achieved because all stakeholders will be more aware of the potential socio-economic trade-offs associated with the increased demand for water used by irrigation to be returned to the rivers for environmental flows.

Irrigation-dependent communities have found it difficult to communicate to the wider populace the benefits of irrigation to their regions and the extent of the initiatives and investment they have undertaken to become more sustainable. As one strategy to improve broader community support for the irrigation sector, irrigation water supply businesses are adopting a broader reporting structure that includes financial, environmental, social and cultural elements. This triple bottom line (TBL) reporting is a holistic approach intended to provide a more sympathetic view of the socio-economic benefits and environmental consequences of water use. It also is intended to provide a useful basis to evaluate the sustainability performance of a business and promote continual improvement.

Triple bottom line reporting by water providers in Australia

VicWater (2002) asserts that the triple bottom line approach is a natural fit with the potable and raw water industry because water is a precious resource and its management by water businesses impacts on the environment, economy and health and well-being of communities. These links are plain enough to see in irrigation-dependent communities during periods of water shortage.

Triple bottom line reporting is not new to the water industry. Australian water providers such as City West Water, Melbourne Water and Sydney Water undertake sustainability reporting, which is considered best practice on the world stage. Although TBL reporting is more prevalent among potable supply businesses, there is increasing incidence of TBL approaches in rural or irrigation water supply organisations. Murray Irrigation Limited (MIL) used the Global Reporting Initiative (GRI) framework to produce an environment report in 2004 that shifts the company reporting towards a TBL performance approach (Murray Irrigation Ltd 2004). Coleambally Irrigation Cooperative Ltd (2004), Murrumbidgee Irrigation Ltd (2004) and Goulburn-Murray Water (2004) adopt aspects of TBL reporting, but all fall short of a GRI compliant sustainability report (SustainAbility 2004). A summary of analysis of each company's report is given in Table 5.1. The table was prepared on the basis of the reports available in 2006.

A number of the water provider organisations' reports we reviewed did not meet the seven criteria for a GRI-compliant sustainability report. The main areas for improvement in reporting are: vision and strategy for sustainability; definition and prioritisation of sustainability challenges for the organisation; presentation or analysis of data on performance across the triple bottom line; and use of assurance processes, including stakeholder comment. We present below two case studies of voluntary public reporting from leading water utilities. The original analyses were conducted in 2006, but have been updated by a review of more recent reporting. They demonstrate the drivers of, and complexities associated with, voluntary public reporting. They also illustrate the volatility of this type of reporting with a lack of well-developed standards, systems and strategies.

Triple bottom line reporting case study: Murray Irrigation Ltd

Murray Irrigation Ltd (MIL) is Australia's largest private irrigation company, formed in 1995 when the NSW Government transferred ownership of the Murray Irrigation Area and Districts to irrigators. As a condition of privatisation, MIL was issued two licences:

- Irrigation Corporation Water Management Works Licence, by the Department of Infrastructure, Planning and Natural Resources
- Environment Protection Licence, by the Department of Environment and Conservation.

MIL also undertook to manage the natural resources of the region and became the implementation authority for the Murray Land and Water Management Plans. These licences and Land and Water Management Plan (LWMP) funding authorities required the production of an annual report to monitor MIL's licence compliance. In part, this was to ensure the company did not neglect their environmental responsibilities and to track progress of LWMP implementation.

Table 5.1: Analysis of sustainability reporting by irrigation industry organisations

Organisation name	Report type	All elements of a sustainability report are met	Sustainability report assessment criteria						
			The report includes elements of TBL reporting	The company presents a coherent vision of sustainability	Sustainability challenges are clearly stated and prioritised by the company	Sustainability strategy is clear by the company	There is a balance of TBL performance data presented	The report is comprehensive, useful and clear in information and design	The report uses various forms of assurance (stakeholder comments, verification, external reviews)
Coleambally Irrigation Co-op Ltd	Annual Report (2003/04)	No	✓	✗	✗	✗	✗	✓	✗
Murray Irrigation Ltd	Environment Report 2004	Yes*	✓	✓	✓	✓	✓*	✓	✓
Murrumbidgee Irrigation Ltd	Annual Report (2003/04)	No	✓	✗	✗	✗	✗	✓	✗
Goulburn-Murray Water	Annual Report (2003/04)	No	✓	✗	✗	✗	✗	✓	✓
North Burdekin Water Board	Annual Report (2003/04)	No	✗	✗	✗	✗	✗	✓	✗
State Water	Annual Report (2003)	No	✓	✗	✓	✓	✗	✓	✗
Sydney Water	Annual Report (2003/04)	Yes	✓	✓	✓	✓	✓	✓	✓
SunWater	Annual Report (2003/04)	Yes	✓	✓	✓	✓	✓	✓	✓
SA Water	Sustainability Report 2003	Yes	✓	✓	✓	✓	✓	✓	✓
City West Water	Sustainability Report 2004	Yes	✓	✓	✓	✓	✓	✓	✓
Melbourne Water	Annual Report (2003/04)	Yes	✓	✓	✓	✓	✓	✓	✓

* Lack of economic information because the financial statements in the Annual Report were not audited at time of publication

Compliance reporting originally involved only the presentation of the required data, but gradually expanded to provide greater context, explanation of trends and beyond compliance activities. By 2000, the report had become large and cumbersome. Although it incorporated 'good news' stories about the region's environmental performance, it was too detailed for a general audience and was not an objective and comprehensive TBL sustainability report.

In a desire to have a better reporting methodology and to demonstrate to the community that irrigated agriculture can be sustainable, various reporting methodologies were researched by MIL. The GRI framework (GRI 2002) provided the company with a logical process to move to more useful to TBL reporting. Once environmental reporting was no longer driven by the licensing, it provided an opportunity to use reports for recognition of good environmental performance, transparency in reporting and enhancement of the credibility and professional standing of the company.

MIL worked towards full compliance with the GRI standards, believing that adopting a reporting initiative such as the GRI would progress improvements in environmental/sustainability reporting more rapidly than developing a new in-house framework. MIL have acknowledged in discussions that achieving GRI status was not as simple as finding the information to fulfil the criteria: systems had to be developed to capture the appropriate data. Perhaps more importantly, the emphasis of GRI on the need for strategies with clear objectives and targets led to serious reflection by the company on its future direction. This led to the development of a 15-year environmental strategy and a business plan that sets targets to report against, also making GRI alignment easier. The GRI has been a significant catalyst to changes in environmental strategies implemented by the corporation.

In an effort to get feedback on their reports, MIL submitted their report to a number of reporting awards such as the Australasian Reporting Awards and Association of Chartered Certified Accountants awards. The 2003/04 Environment report received a bronze award. In 2006, they included a postage paid feedback card in their report to gauge the interest and usefulness of different sections to the reader.

By 2007, MIL were still reporting against the GRI index and that their reporting was in alignment with the Department of Environment and Heritage Environmental Reporting Guidelines (Department of Environment and Heritage 2002) and had largely incorporated the National Water Commission's 2007 National Performance Framework for rural water providers (National Water Commission 2008). However, since 2007, MIL annual reports no longer report against the GRI index. They describe reporting as reflecting overall company performance in accordance with its strategic plan and statutory reporting requirements. This move away from explicit GRI style reporting has meant that reporting on recycling activities, staff gender profiles and other indices less associated with governance and compliance reporting has declined. This has resulted in about a 25% reduction in the length of the report. It is possible to interpret this move away from a focus on the GRI as the standard for social and environmental performance either as a form of de-emphasis or as reflecting a stronger integration of social and environmental considerations into the heart of the corporation's strategic thinking. In either case, there has been a shift away from an emphasis on the documentation as the central concern towards a focus on sustainability and strategy, and upon integration of all aspects of corporate reporting.

Triple bottom line reporting case study: Goulburn-Murray Water

Goulburn-Murray Water (G-MW) was formed as a Rural Water Authority in 1994. It was established as Goulburn-Murray Water Corporation as of 1 July 2007. It operates under the *Water Act 1989* (Vic), and is responsible to the Minister for Water. Its primary operations are managing water-related services in a region of 68 000 km^2, bordered by the Great Dividing

Range in the south to the River Murray in the north, and stretching from Corryong in the east, downriver to Nyah. As part of its 'Statement of Obligations' it is required to 'effectively integrate economic, environmental and social objectives into its business operations' (Victorian Government 2009).

In its 2004/2005 Annual Report, G-MW moved to an explicitly TBL structure. As part of a drive for continuous improvement and enhanced sustainability, it took part in a project to review its annual reports against international standards and identify any areas for improvement. In 2007, a series of interviews with G-MW staff were conducted to look at reporting costs and possibilities for improving the organisation's TBL reporting.

Some areas were identified where G-MW's activities had social and environmental effects, but which fell outside its formal reporting responsibilities. These included the social and environmental effects of dams, work on environmental projects supported by Catchment Management Authorities (CMAs) and other bodies, and the effects from water trading. Although it was generally felt that reporting of these would be desirable, in many cases there were significant barriers to obtaining the relevant information. The necessary social data might be obtainable through the multiple local councils in the region, but most likely this would require extensive surveys of the relevant communities and businesses. Environmental effects may be long term, occur outside G-MW's region of operations, and/or be currently unknown, limiting the usefulness of such surveys as a reflection of the actions and impacts of the authority.

There are good reasons for G-MW to be reluctant to commit resources to overcome the barriers to reporting broader TBL information. The general picture of G-MW reporting was of a gradual, but consistent, increase in the extent of reporting required by outside, mostly government, bodies. The corresponding increase in costs creates a pressure to reduce reporting expenses rather than increase them. Also, the view was expressed that monitoring and reporting on these issues might be more appropriately done at a state level.

By the 2008/2009 Annual Report, G-MW had moved away from a TBL structure and, although the report still contains information on initiatives with social and environmental benefits, none of the areas identified in the interviews has been integrated into its annual report.

TBL tools for reporting in the irrigation sector

Both case studies demonstrate some common features. Both organisations, in common with many rural or natural resource dependent industries, have seen the value of public reporting as an element in their strategies to defend and enhance their social licence. Each has looked for an appropriate, independent reporting standard and framework, and the GRI has seemed to offer this. However, both have found through experience that the absence of an integrated strategic framework, the lack of meaningful and accessible data, and the absence of a solid business purposes justification for GRI-compliant reporting have triggered consideration.

For the irrigation industry, there does not exist an entirely appropriate framework for TBL reporting. The GRI has been touted as an option and a number of irrigation companies have incorporated elements. The GRI clearly identifies both the strengths and limitations of their system, and emphasises the need to use the GRI guidelines with other sustainability assessment methods to overcome the limitations in applying it to complex systems. The GRI also proposes a highly generalised three-tier structuring system, comprising categories, aspects and indicators. However, in dealing with complex systems, a more sector-focused and objective-driven structure may be needed to deal more comprehensively with the sector-specific sustainability issues (Chesson *et al.* 2000).

The GRI approach, in a highly modified form, has been adopted, or at least elements incorporated, by urban and rural water providers in Australia, because it is a fairly flexible and adaptable framework. However, these rural water providers have found that the secondary or tertiary industry bias does not make it entirely suitable for primary industries such as irrigation. Most have moved to other approaches.

The relatively straightforward approach of using a selected core set of indicators has been widely used for sustainability assessment and reporting in the irrigation sector. However, in dealing with complex natural resource–socio-economic systems such as agriculture and fisheries, selected core sets of indicators have failed to live up to expectations (Chesson 2002). A more successful approach has been to use a structured approach to identify the main issues of concern to stakeholders or the objectives relating to sustainability, and then consider these objectives using selected indicators and performance measures (Chesson *et al.* 2000). This correctly shifts the focus to what the stakeholders want to achieve. The indicators become a means of reporting against the specific objectives, which are clearly relevant to those stakeholders. This approach seems to be more widely adopted now, with water suppliers reporting less on specific indicators and more on outcomes regarding specific objectives or projects within their business.

Christen *et al.* (2006), developed an *Irrigation Sustainability Assessment Framework* (ISAF), adopting a structured objective-driven approach aimed at the irrigation water provider and urban water provider companies. However, the uptake of this approach has been limited, though some city councils in Sydney have used the approach for urban irrigation water management. In Chapter 13 of this book, Stone and Martin outline a complementary approach to linking voluntary reporting and board-level strategic decisions and community engagement.

Conclusions

The use of irrigation in inland south-eastern Australia has resulted in productive systems that support a population and revenue base that would not be possible from rain-dependent agriculture alone. In the process of bringing about this development, there has been considerable change in the land and vegetation and the supplying river systems. The current wider community assessment is that the net benefit from irrigation and other land use change is too heavily weighted to production outcomes at too great an expense of the environmental resources. The social licence of irrigators is being heavily contested in the media and the political arena, and the consequences of these debates are strategically important. Social licence is no longer a term of only academic relevance to the individual primary producer. These contests over water are not the only aspect of farming where social licence is operationally important: animal welfare, regulation of landscape conservation, debates about the incidence of carbon pricing on farming are all illustrative. There is considerable pressure on irrigators and the water supply authorities to improve their performance and to demonstrate their socially beneficial effects, not only in the economic dimension but also in the social and environmental dimensions. The use of a TBL management and reporting arrangement is one way of identifying where improvement can be made and gauging how this is changing over time. However, the practical challenges remain enormous.

The TBL provides a dual function as a tool for business management planning and a framework for reporting that places business in the context of widely accepted approaches to sustainability within society. There is a desire by the community that businesses/organisations become more responsible and transparent about the social sustainability issues over which they have influence or impact. This desire for an ethical and accountable approach to business management links with the drivers of financial value, risk management, compliance with legislation and benchmarking performance in sustainability reporting.

Irrigation organisations are continuously setting TBL sustainability objectives, developing management goals and reporting performance on achieving these goals. At this time, there is no overarching reporting framework for the irrigation sector, so organisations are following various pathways.

The impact of social licence has been to make irrigation water supply businesses more accountable to community expectations. However, there is a risk that this will only lead to a fruitless pursuit to meet ever-shifting community concerns. This chapter has highlighted that for a business to successfully engage with its social licence demands an approach to performance management, run by the business and at the heart of its strategic planning. It has been demonstrated that while the need has been recognised by water utilities, the absence of integrated reporting frameworks, which are logically connected to strategy and operations, is a significant practical impediment. Coupled with this are significant gaps in the data that can be used to construct reliable social and ecological reports, at levels that are relevant to the communities to whom these corporations feel they ought be most accountable. Social reporting is an important component of the social licence strategies of primary industries, but substantial development and investment will be needed before the potential can be fully realised.

References and further reading

Blackburn G (2004) *Pioneering Irrigation in Australia to 1920.* Australian Scholarly Publishing Pty Ltd, Kew, Victoria.

Brundtland G (Ed.) (1987) *Our Common Future: The World Commission on Environment and Development.* Oxford University Press, Oxford.

Centre for Australian Ethical Research (2004) 'The state of sustainability reporting in Australia'. Australian Government Department of Environment and Heritage, Canberra.

Chesson D (2002) 'Sustainability indicators – measuring our progress'. Bureau of Rural Science, Canberra.

Chesson D, Whitworth B and Smith T (2000) 'Reporting on ecologically sustainable development: the reporting framework of the standing committee on fisheries and aquaculture in relation to national and international experience'. Final report to Fisheries Resources Research Fund, December 2000. Bureau of Rural Science, Canberra.

Christen EW, Shepheard ML, Meyer WS, Jayawardane NS and Fairweather H (2006) Triple bottom line reporting to promote sustainability of irrigation in Australia. *Journal of Irrigation and Drainage Systems* **20**(4), 329–343.

Coleambally Irrigation Co-operative Limited (2004) 'Annual report 2003–2004'. Coleambally Irrigation Co-operative Limited, Coleambally, New South Wales.

Commonwealth of Australia (1992) 'National strategy for ecologically sustainable development'. Australian Government Publishing Service, Canberra.

Department of Environment and Heritage (2002) 'Department of Environment and Heritage Environmental Reporting Guidelines'. Department of Environment and Heritage, Canberra.

Elkington J (1998) *Cannibals with Forks: The Triple Bottom Line of 21st Century Business.* 2nd edn. New Society Publishers, Gabriola Island, Canada.

Goulburn-Murray Water (2004) 'Annual report 2003/04'. Goulburn Murray Water, Tatura, Victoria.

Global Reporting Initiative (2002) 'Sustainability reporting guidelines'. Global Reporting Initiative, Amsterdam, The Netherlands.

Hallows PJ and Thompson DG (1995) 'The history of irrigation in Australia'. Australian National Committee on Irrigation and Drainage, Mildura.

Mack D (2003) 'Irrigation settlement. Some historic aspects in South Australia on the River Murray 1838–1978'. Cobdogla Irrigation and Steam Museum, Berri, South Australia.

Meyer WS (2005) 'The irrigation industry in the Murray and Murrumbidgee basins'. CRC Irrigation Futures Technical Report No. 03/05, April 2005, CRC Irrigation Futures, Richmond.

Murrumbidgee Irrigation Ltd (2004) 'Annual report 2003–2004'. Murrumbidgee Irrigation Ltd, Griffith, New South Wales.

Murray Irrigation Limited (2004) 'Murray irrigation environment report 2004'. Murray Irrigation Ltd, Deniliquin, New South Wales.

National Water Commission (2004) 'The national water initiative'. <http://www.nwc.gov.au> (accessed 18 July 2005).

National Water Commission (2008) 'National performance report 2006–07: Rural Water providers'. <http://www.nwc.gov.au> (accessed 21 March 2009).

Nelson L and Wilson C (2003) *Triple Bottom Line: A New Approach to Reporting Your Organisation's Performance.* Tertiary Press, Croyden, Victoria.

State Water (2004) 'State water annual report 2003–04'. State Water, Sydney.

Suggett D and Goodsir B (2002) 'Triple bottom line measurement and reporting in Australia: making it Tangible'. The Allen Consulting Group, Melbourne.

SunWater (2004) 'SunWater annual report 2003–04'. SunWater, Brisbane.

SustainAbility (2004) 'Risk and opportunity. Best practice in non-financial reporting'. Standard and Poor's and United Nations Environment Program.

Vanclay F (2003) Experiences from the field of social impact assessment: where do TBL, EIA and SIA fit in relation to each other? In: *Social Dimensions of the Triple Bottom Line in Rural Australia.* (Eds B Pritchard, A Curtis, J Spriggs and R Le Heron) pp. 61–80. Australian Government Bureau of Rural Sciences, Canberra.

van der Lee J and Wolfenden J (2002) 'A triple bottom line framework for reporting'. Final report presented to the Cotton Research and Development Corporation. Centre for Ecological Economics and Water Policy Research, University of New England, Armidale.

Victorian Government (2009) 'Statement of obligations'. <http://www.ourwater.vic.gov.au/governance/water-corporations/statement_of_obligations>.

VicWater (2002) 'Triple bottom line reporting guidelines'. The Victorian Water Industry Association, Melbourne.

6

Social licence issues in developing economies

Donna Craig and Michael Jeffery

This chapter considers the concept of a *social licence to farm* from the perspective of India, which has evolved over time from a colonial past into a democracy with a strong legal system and modern constitution. We suggest that India could have the potential to provide the legal framework to significantly advance the cause of sustainable agriculture. Unlike developed countries such as the US, Canada, Australia, Great Britain and France, India's Constitution provides a right to life and explicitly incorporates environmental protection and improvement as part of state policy. It therefore has, to some extent, a structural advantage, although without more progress on poverty alleviation and food security the path forward will be painstakingly slow. History, population pressures, extreme poverty, economic, cultural and political capacity constraints, among many others, are all factors that must be taken into account to understand the inability of both the developed and developing world to move more rapidly towards a more sustainable form of agriculture.

This chapter will focus on the development of agriculture in India, with some comparative examples from pesticide use in China and Bhutan. The problems identified may reflect the grim reality of developing nations, such as India, with relatively few resources to tackle the social licence aspect of farming. We will attempt to provide a different perspective on social licence to farm issues than may emerge from other chapters. We will highlight the competition between industry's desire to use the environment as a 'sink' for noxious by-products and farmers' desire to use the same environment for their economic interests. We will also highlight the problem of farmers themselves as sources of harmful pollutants. In this way, we hope to demonstrate that conflicts over social licence are about the legal and political system rebalancing the private interests of competing users. In the case of India, we will demonstrate how the Constitution, an overarching instruction to the courts about how they ought resolve such tensions, is a critical determinant to the outcome of such social licence to the environment conflicts. The chapter will conclude by briefly highlighting the equally serious problems facing a highly developed country (Australia). We will discuss the unsustainable farming practices and water usage of Cubby Station in the Murray–Darling Basin.

Origins of agriculture

Although surprising for citizens living in the 21st century, it is nevertheless true that human societies during more than 99% of their existence on earth have lived as hunter/gatherers.[1] Before humans started producing food about 10 000 years ago, they lived off the resources provided directly from nature. Human communities entered a new stage of cultural development when, instead of depending entirely on the resources of nature for survival, they started producing food by cultivating cereals and domesticating animals.

The beginning of 'agriculture' marks the beginning of 'civilised' or 'sedentary' society. Climate change and increase in population during the Holocene Era (10 000 BC onwards) led to the evolution of agriculture. During the Bronze Age (9000 BC onwards), domestication of plants and animals changed the profession of the early *Homo sapiens* from hunting and gathering to selective hunting, herding and finally to settled agriculture (Gupta 2004). Eventually, the agricultural practices enabled people to establish permanent settlements and expand urban-based societies. Cultivation marked the transition from nomadic pre-historic societies to the settled Neolithic lifestyle sometime around 7000 BC. Associated with this change in the relationship between humans and nature came fundamental changes in the relationship between people. Law and politics, and advances in technology, became increasingly central to the securing of necessary resources, and economics rather than direct interaction with the natural world became the mechanism for winning wealth. The wealthier and more 'advanced' the community, the more that this is characteristic of the relationship between people and nature.

Agriculture in India

Agriculture in India is not only an important occupation of the people, but also a way of life, culture and custom. Agriculture in India is the pre-eminent sector of the economy. It is the source of livelihood of almost two-thirds of the workforce in the country. The contribution of agriculture and allied activities to India's economic growth in recent years has been no less significant than that of industry and services. Most of the Indian customs and festivals are observed in consonance with agriculture seasons, activities and products. These festivals are celebrated with different names and rituals in almost all the states of India. For instance, Basanat Panchami and Baisakhi are celebrated in North India and Pongal and Onam in South India.

History of agriculture in India

The history of Indian agriculture dates back over 10 000 years. Indian agriculture began during 9000 BCE with the early cultivation of plants, and domestication of crops and animals. The Indus civilisation agriculture was highly productive. It was capable of producing surpluses sufficient to support tens of thousands of urban residents whose primary occupation was not agriculture. Throughout the history of India, agriculture has been its predominant activity. Many contemporary Indian and foreign writers have praised the fertility of Indian soil. Though agriculture mainly depended on rainwater, a variety of other methods of artificial irrigation were also employed. Wells fitted with various devices to pull water, tanks, reservoirs and, to a limited extent, canals were also the source of irrigation.

However, with the advent of the British, many transformations took place in the field of economy, law, administration and other spheres of life. The establishment of colonial rule in India, and various policies of the colonial government, had a devastating effect on the farmers and gave rise to many popular uprisings. The world's worst recorded food disaster happened in 1943 in British India, known as the *Bengal Famine*, where an estimated four million people died of hunger that year alone in eastern India (that included today's Bangladesh). These factors, both historical and contemporary, make issues of agriculture, private rights and social/environmental responsibility matters of great political and legal importance to a large part of India's burgeoning population. Coupled with a robust democratic culture, this means that the social licence of farmers is a mainstream political issue.

Agriculture in independent India

After independence in 1947, India continued to be haunted by memories of the Bengal Famine. It was therefore natural that food security was paramount on the agenda of the free India. The early years of Independence witnessed a marked emphasis on the development of infrastructure for scientific agriculture. The steps taken included: the establishment of fertiliser and pesticide factories; construction of large multi-purpose irrigation/power projects; and the organisation of community development and national extension programs. However, the growth in food production was inadequate to meet the consumption needs of the growing population. This necessitated food imports.

The Green Revolution, spreading over the period from 1967 to 1978, introduced use of high-yielding varieties of seeds and increased use of fertilisers, pesticides and irrigation. By the early 1990s, India's status changed from a food-deficient country to one of the world's leading agricultural nations. The states of Punjab and Haryana emerged as the food bowls of India. In spite of this, India's agricultural output often falls short of demand even today. The Green Revolution showed best results only in the states of Punjab and Haryana. The results were less impressive in other parts of India. Not everyone has access to food, and India is still the country with the poorest people worldwide.

Indian agriculture continues to face internal and external challenges. Monsoon dependence, fragmented land-holding, low levels of input usage, antiquated agronomic practices, and lack of technology application and poor rural infrastructure are some of the key internal constraints that deter a healthy growth, while subsidies and barriers have been distorting international agricultural trade, rendering agri-exports uncompetitive. Such circumstances are far from unique to India, being shared by many other developing countries.

The objective of virtually every policy initiative has been to make Indian agriculture globally competitive – by investing it with the ability to produce globally acceptable quality at globally comparable cost. India is urbanising, but 73% of the population still lives in the rural India. Most people in rural India depend directly or indirectly on farming for their livelihood. Despite this, not much attention has been given to agriculture to overcome poverty. A rising population and demands for increased standard of living puts more stress on agricultural practices. National farm output must rise if these demands are to be met.

India is a diverse, vast country with 28 states and seven Union Territories that differ vastly in terms of their natural resources, administrative capacity and economic performance (Sustainet 2006). Some states, especially northern and north-eastern states remain very poor and the farmers continue to be the most exploited and oppressed class all over India. Most of the farms are small scale: about half of all farms are less than 1 hectare in size, and another 20% are less than 2 hectares (Sustainet 2006).

Recently there have been media accounts about suicides among farmers from all over the country, particularly from states of Andhra Pradesh, Maharashtra; Karnataka, Kerala and Punjab. The situation was grim enough to force at least the Maharashtra government to set up a dedicated office to deal with farmers' distress (Meeta and Rajivlochan 2006). One Telugu newspaper reported as many as 349 deaths within 6 months in the state of Andhra Pradesh.[2] According to Sainath (2009), between 1997 and 2007, a staggering 182 936 farmers committed suicide. The government has taken many initiatives to help the farmers, for example, by providing subsidies. It is possible to hear some echoes of the issues affecting farmers in more developed economies in such statistics, but with a far greater degree of desperation and need. The legal system in India has responded to this greater criticality of issues with institutional arrangements that are quite different to those in wealthy, less farm-dependent societies.

The following sections provide an overview of the constitutional and legal frameworks that have potential to deal with issues related to social licence to farm, including the protection of human and ecosystem health. Importantly, some examples are given of the role of the Indian judiciary in providing redress when laws and policies are poorly, or unfairly, implemented.

Article 21 of the Indian Constitution states:

'No person shall be deprived of his life or personal liberty except according to procedures established by law.'

The Supreme Court expanded this negative right in two ways. Firstly, any law affecting personal liberty should be reasonable, fair and just.[3] Secondly, the Court recognised several unarticulated liberties[4] that were implied by Article 21 and thus interpreted the right to life and personal liberty to include the right to a clean environment.[5]

In addition, the *Constitution (Forty Second Amendment) Act 1976* explicitly incorporated environmental protection and improvement as a part of state policy. Article 48A, a Directive Principle of State Policy, provides that:

'The State shall endeavour to protect and improve the environment and safeguard the forests and wildlife of the country.'

Moreover, article 51A (g) imposes a similar responsibility on every citizen:

'to protect and improve the natural environment including forests, lakes, rivers and wildlife, and to have compassion for living creatures ...'

Therefore, protection of natural environment and compassion for living creatures were made the positive fundamental duty of every citizen. Both the provisions substantially send the same message. Together, they highlight the national consensus on the importance of the protection and improvement of the environment.

India's legal regulatory framework

Food safety issues have proven to be particularly challenging in India, in part because of the advent of modern chemical-intensive farming in a society where farmer and institutional sophistication lags that which exists in the economies where such productive innovations originate.

Chemical poisoning and adulteration of food crops emerged alongside the 'green revolution'. In order to curb problems associated with food adulteration, the Government of India introduced the *Prevention of Food Adulteration Act 1954* along with the *Prevention of Food Adulteration Rules 1955*. Unfortunately, they did not prove to be very effective in curbing this crime.

The Parliament of India enacted the *Food Safety and Standards Act* in 2006 to consolidate laws relating to food and ensures wholesome food for human consumption. The *Insecticides Act 1968* regulates the import, manufacture, sale, transport, distribution and use of insecticides with a view to prevent risk to human beings or animals.

It is unfortunate that despite the constitutional emphasis on both the health and safety of Indian citizens, and the natural environment, indiscriminate use of insecticides and pesticides is rampant in India. Poor regulation and low levels of education and awareness in the farming population contributes to farmers and their dependents being constantly exposed to the harmful effects of these chemicals and suffering many ailments.

With the problems associated with chemical pesticides becoming more evident, the industry has embraced a relatively new technology in the form of insect-resistant genetically engineered crops such as 'Bt cotton', which are portrayed as a panacea for controlling pests. However, during the last few years of commercial cultivation of 'Bt cotton', especially in the State of Andhra Pradesh, the devastating effects such technologies can have on farming communities has become apparent (Sustainet 2006). 'Bt cotton' seed is four times the price of conventional seeds and 'Bt' crops often are not completely resistant to the pests that they are designed to combat. Other pests can still attack the crops, requiring the continued use of chemical pesticides. The indiscriminate use and frequent high doses of pesticides has been held responsible for major 'pest disasters', such as the many suicides of farmers in Andhra Pradesh in 1997–98 (Sustainet 2006). In Chapter 7 in this book, the ecological and economic benefits of Bt cotton and associated herbicides are discussed. It is salutary to note that the same technologies, used in different economic and social settings, seem to have had markedly different effects. This suggests that matters of social licence can never be considered without a sophisticated awareness of the social context. What may be defensible or even beneficial in one context can be hazardous in another.

Case studies – the courts intervene

Backed by the Indian constitution, farmers have been able to have recourse to the courts to defend key aspects of their livelihoods. Three examples, all associated with water, serve to illustrate.

(1) Village of Bichhri – State of Rajasthan

There are numerous examples of cases being filed on behalf of farmers for redressing their grievances. One such case was filed by the NGO, Indian Council for Enviro-Legal Action in the Supreme Court regarding groundwater pollution in Bichhri.[6]

In 1987, the Hindustan Agro Ltd started producing chemicals such as oleum (a concentrated form of sulphuric acid) and single superphosphate. A second company, Silver Chemical, commenced production of hydrochloric acid in the same complex. The production of hydrochloric acid led to the generation of large quantities of highly toxic effluents and iron and gypsum sludge, causing damage to the land. These industries disposed of large quantities of sludge in the open, in and around the complex. The leachate from toxic sludge percolated deep into the ground polluting the aquifers in the drought-prone area, further constraining the limited supply of water.

The water in the wells turned a dark colour, like black coffee, and became unfit for human consumption. The water also became unfit for cattle and for irrigation. The soil was contaminated and unfit for cultivation. Death and economic disaster in the village and surrounding areas were reported. With only the annual monsoon to depend on, the largely agricultural community of Bichhri was able to produce just one crop annually. Over 70 wells in 22 villages (Mehta 2009) in and around Bichhri began pumping up poison. Land, which earlier yielded up to 10 bags of rice per *bigha* (an Indian unit of measurement, which varies from 1.5 per acre to 3.7 per acre), was now producing only about two bags (Sharma and Banerji 1996).

With the polluted wells, the community effectively lost their only source of irrigation for their agricultural land. They also developed breathing problems, skin lesions and other disorders. Though the companies responsible for the pollution adopted every means to avoid being blamed for the contamination and paying compensation to the victims, the court's historic judgement (on 13 February 1996) established the *polluter pays principle* in the context of Indian environmental law holding the companies accountable for the loss and for paying the costs of remediation.[7]

The court found that all of the regulatory agencies, including the Central Government, had failed to force the polluter to pay. The court, in rendering its judgement, accepted the validity of the absolute liability principle in the case.[8] Although in contemporary advanced economies this may not be seen as a surprising outcome, in the case of India in the early 1990s, this outcome is significant. At a time when the country was aggressively pursuing industrialisation (some would argue, at almost any cost), farmers were able to secure a significant and effective protection of their licence to use water, over the interests of an economically important industry.

(2) Vellore – State of Tamil Nadu

In another case, in Vellore, approximately 600 tanneries discharged untreated effluent into agricultural fields, road-sides, waterways and open lands. The untreated waste was finally discharged in the Palar River, the main source of water supply to the residents of the area. It became polluted to such an extent that the whole river became unusable within a few years. Women and children walked miles to get drinking water. All of the agricultural land in the 'Tanneries Belt' became unfit for cultivation. The farmers effectively lost their fertile agricultural land and principal source of water. Once again, the Supreme Court came to their rescue in *Vellore Citizens Welfare Forum v Union of India* and required the tanneries to pay for pollution (Judgement dated 28 August 1996).[9] The Court in a scathing judgement stated:

> *'It is high time that the Central Government realises its responsibility and statutory duty to protect the degrading environment in the country. The tanneries of the State Tamil Nadu, India will compensate affected people and will pay the Union of India to complete environmental restoration.'*[10]

The Court also ordered that:

- the Central Government create an Authority to deal with the situation created by the tanneries and other polluting industries
- the Authority must compute the compensation for reversing the ecological harm and for compensation to individuals
- tanneries must meet the formal environmental standards and stipulations from henceforward
- an Agreement drawn up to put in place the necessary pollution control devices does not fee industry from an obligation to pay for the past pollution generated
- tanneries must pay a pollution fine of Rs.10 000/– each: The money, along with the compensation amount recovered from the polluters, was to be put into an 'Environment Protection Fund' to be used for compensating the affected persons and for restoring the damaged environment
- The State Government, under the supervision of the Central Government, was to execute restoration. The cost will be funded by the 'Environment Protection Fund' and from other sources provided by the State Government and the Central Government.[11]

Once again, the combination of a clear constitutional mandate for protection of the environment and a diligent legal institution provided effective redress for the affected farmers.

(3) Commercial farming – Tamil Nadu

In yet another case in the State of Tamil Nadu, large industrial houses and multi-nationals started large-scale commercial farming in an unplanned and unscientific manner. These aquaculture businesses destroyed large areas of rich fertility, made the groundwater saline and destroyed the livelihood of fishermen in land that for centuries had been richly covered with rice paddies.

Prawns were grown in large artificial tanks. Prawn growth is best when the water in these tanks is brackish (part fresh and part salt). The fresh water was taken from the ground, depleting the groundwater supply, while the saltwater was brought from the sea, necessitating the cutting down of the coast-protecting mangrove forests, which are the breeding grounds of fish. The salt water, contaminated with hormone-containing chemical foods and pesticides used in the tanks, seeped into the groundwater. This made well-water undrinkable and led to the deaths of cattle and skin and eye diseases among humans.

The polluted water also drained into the rivers and the ocean. Fish catches reduced by as much as 80%[12] in some places because of surface water contamination and mangrove deforestation. The saltwater seeped into the cropland, making it barren. The case was filed in the Indian Supreme Court, which ordered the shutdown of almost all the prawn farms in Tamil Nadu.[13] An Authority was constituted by the Central Government under *Environment Protection Act 1986* to implement the 'Precautionary and Polluter Pays' principles to assess the loss to ecology and recover the cost of eco-restoration and amount of compensation from the polluters.[14]

What do these examples illustrate about social licence and farming? It is interesting to note that in every case protection of the effective freedom to farm necessarily involved curtailing the 'freedom' of another set of resource users to pursue their own economic interests. This highlights an important characteristic of social licence competition. Frequently, it is not enough to consider the impacts of restriction on one interest group alone. Under conditions of resource scarcity, adjusting one set of interests intrinsically means adjusting another in the opposite direction. In the examples given, these adjustments have re-established the interests of farmers, whereas in other situations the effect will be to reduce the interests of farmers.

In India, as in the rest of the world, competition for resource continues to intensify. Re-adjustment of exploitative (or even conserving) interests will almost always be a win:lose political contest. What is different about India is the existence of a constitutional framework, which nudges the Court system towards protection of the interest of both the environment and the small-scale farmers.

Pesticide use in China and Bhutan

The problem of pollution is not very different in other developing countries such as China. China is the world's biggest user, producer and exporter of pesticides (Yang Yang 2010). The growing use of pesticides and chemical fertilisers has helped promote larger crop yields, but increasingly at a major cost to the environment and human health. Once again, the interests of farmers do not necessarily align with other interests, and the law is involved. A survey conducted by Greenpeace China on vegetables sold in China's large cities found many pesticide-ridden vegetables, some of them even carrying cocktails of many highly poisonous pesticides (Shan 2010). Among the pesticides found, five of them were known for causing cancer.

China has received increased international media scrutiny in recent years following political reforms, joining the World Trade Organisation and the increased awareness of Chinese people in urban areas of food safety. There have been numerous incidents involving food safety in the People's Republic of China (PRC). These have included the unconventional use of pesticides or other dangerous chemical additives, such as food preservatives or additives, and the use of unhygienic starting materials as food ingredients. The 2008 Chinese milk scandal resulting in the deaths of several children received the most attention among food safety incidents.

The consumption of pesticides in China has increased during the last few years. More than 1 million tonnes of finished pesticide products are spread on agricultural land (Linthoingambi *et al.* 2009). The Chinese Government figures show that 53 300 to 123 000 people are poisoned by pesticides each year (Linthoingambi *et al.* 2009). Many Chinese farmers are not willing, or are not able, to invest in protective clothing and equipment for safe pesticide use, which has greatly increased the risk of pesticide poisoning. Hundreds of farmers die each year due to improper use of pesticides. Many farmers suffer liver, kidney, nerve and blood problems due to pesticide poisoning, as well as eye problems, headaches, skin effects and respiratory irritations (Linthoingambi *et al.* 2009). We suspect that if they were asked whether the government ought 'interfere' by regulation in their farming practices, including the use of chemicals, herbicides and pesticides, Chinese farmers, like farmers around the world, would protest that this is an unjustified interference with their social licence to farm.

The manufacture and sale of agricultural pesticides in China is regulated by the Institute for the Control of Agrochemicals of the Ministry of Agriculture (ICAMA), which was established in 1963 by the Ministry of Agriculture (MOA). In addition to ICAMA's regulation of pesticides at the national level, pesticides are regulated under a federated program at the province, city and county levels. ICAMA's responsibilities include: implementing regulations; issuing pesticide registration guidelines; reviewing and approving pesticide advertising; inspecting markets that sell pesticides; supervising pesticide quality, conducting efficacy and residue trials and training technicians and administrators of the provinces.

China's principal law governing pesticides, the *Regulation on Pesticide Administration* (RPA) was issued on May 8, 1997 by the State Council and amended on November 29, 2001. On 27 April 1999, MOA adopted Implementations for RPA and amended these regulations in July 2002 and 2004. In 2001, MOA adopted 'Data Requirements of the Pesticide Registration'. ICAMA has also issued regulations on product quality and pesticide advertising. In recent years, the Chinese Government attempted to consolidate food safety regulation with the creation of the 'State Food and Drug Administration of China' in 2003. In October 2007, China approved new legislation aimed at improving and monitoring national standards in food production. New laws will standardise food production and clamp down on illegal activity in the farming and farm chemicals industries.

The extent to which legal regulation is needed to protect farmers, consumers and the environment varies significantly with the social and technological context. We have already suggested this in relation to genetically modified cotton. This same point can be made in contrasting pesticide and chemical use in China, with the same issues in Bhutan: a small country in South Asia in the Himalayas. China's rate of adoption of modern farming technology has been far more rapid than has been the case for Bhutan. One result has been that, while Bhutan has lagged China in production improvement, it has also thus far avoided the worst of the problems that have harmed China's citizens and their environment.

This suggests the possibility that one important consideration in managing social licence conflicts is that the legal institutions for managing these issues evolve in parallel with the emergence of resource demands.

Similar to India, in Bhutan, environmental conservation has been recognised as a Constitutional mandate, which stipulates that the Bhutan Government, and every Bhutanese, is a trustee of the Kingdom's natural resources and environment for the benefit of the present and future generations.[15] It is the fundamental duty of every citizen to contribute to the protection of the natural environment, conservation of the rich biodiversity of Bhutan and prevention of all forms of ecological degradation including noise, visual and physical pollution through the adoption and support of environment friendly practices and policies.[16] Bhutan is the first country in the world that began the concept of 'gross national happiness' (GNH), first propounded by His Majesty the Fourth Druk Gyalpo Jigme Singye Wangchuck in the 1980s. Although conventional development models stress economic growth as the ultimate objective, the concept of GNH is based on the premise that true development of human society takes place when material, spiritual and emotional well-being occur side by side to complement and reinforce each other.[17]

Bhutan is facing acute land degradation due to: forest fires; excessive use of forest resources; overgrazing; unsustainable agricultural practices; poor irrigation system management; construction of infrastructure without proper environmental measures; mining; industrial development; and urbanisation.[18] Traditionally, the farmers relied on farmyard manure (cattle dung, etc.) as fertiliser, but, in the 1960s, use of inorganic fertilisers became dominant. Although the absolute levels of use of inorganic fertilisers in Bhutan is still comparatively low by global standards, at the household level use has becoming increasingly significant. The use of fertilisers in a disproportionate manner, such as using urea (a nitrogen-supplying compound) is common because it is affordable compared with other inorganic fertilisers. This results in an imbalance in soil nutrient management. The use of chemical pesticides is seen as an easy, inexpensive and quick solution for controlling insect pests and weeds. However, improper and prolonged use of chemical pesticides can pose significant environmental risks, such as contamination of land and water (both ground and surface) and attrition of non-target organisms, ranging from beneficial soil microorganisms to fish and birds. Bhutan has demonstrated a radically different approach to balancing its social and economic interests to the rest of the world. It has embedded a commitment to social justice and resource sustainability, akin to India, but with a more fundamental national will. The cost, in terms of economic growth, has been high, but time will tell if this has been a sensible choice. Bhutanese farmers are taking small steps down the same technological paths that have delivered such mixed blessings as have been seen in China and India.

The key to whether Bhutan will be able to navigate the path between maximum productivity and ecological sustainability more effectively than other developing nations will be its culture, laws and institutions. One essential element will be how these formal structures manage to restrict and channel the freedoms of individual farmers, and how the pace of modernisation is adjusted to the capacity of society to increase the productivity without incurring excessive social and ecological costs.

Social licence and water security

From the perspective of having considered social licence issues in some quite different circumstances to those that prevail in the more developed world, we will turn briefly to looking at a similar type of challenge in Australia.

The global situation regarding water security is severe and the way we treat water resources is a cause of concern. The world is already facing a water crisis. According to a WWF report,[19] over 2 billion people are affected by water shortages in over 40 countries. The extensive withdrawal of water for agriculture from river, lakes and aquifers results in limited supplies for other human needs, such as drinking, washing, cooking and sanitation. According to the UN World Water Development Report,[20] the average supply of water per person will drop by a third in the next two decades and the expected global water withdrawal for irrigation will rise 14% higher.

Inappropriate irrigation practice, or growing crops not suitable to the local environment, leads to wastage of much of the water used for irrigation. It is argued that waste is driven by misplaced subsidies, artificially low water prices (usually not connected with the amount used), low public and political awareness, poor water management and inadequate environmental legislation. Unsustainable agriculture harms the environment by draining rivers, lakes and depletes underground water sources, increases soil salinity and washes pollutants and pesticides into rivers. Ultimately, this can destroy the livelihoods of the farmers practising it, and of fishermen and communities dependent on natural ecosystems.

The UN Report argues that agriculture withdraws the vast majority of water taken from rivers, lakes and underground aquifers. Making the required amount of water available is only possible, through the damming and diversion of rivers and extensive pumping of underground aquifers. The report further says that agriculture is the largest source of pollution, including water pollution, in most countries.[21]

Widespread misuse of groundwater by farmers is a serious concern in India and Bhutan. The groundwater reserves form one of the most important sources of irrigation. In India, about 60% of the irrigated food grain production depends on groundwater and about half of total area irrigated depends on groundwater wells.[22] Due to indiscriminate and large-scale extraction of this valuable resource, the rate of use far exceeds the rate of recharge. The level of groundwater is depleting alarmingly all over the country.

In most cases, the farmers are provided electricity at subsidised rates and are oblivious to the needless wastage of electricity. Similarly, there is widespread wastage of groundwater through improper use of pumps and tube wells. This results in a huge cost to society and further depletes water reserves. Though the Government of India constituted the Central Ground Water Authority (CGWA) with an aim to regulate the indiscriminate boring and withdrawal of groundwater in the country, it has not proven effective. Indian law relating to groundwater allows the owner of a property unlimited use of the groundwater. Increased reliance on groundwater in the future will allow continued depletion of the resource, if not managed scientifically and properly regulated. This situation suggests that India will soon face the necessity of 'clawing back' rights and entitlements to water upon which farmers have come to depend. Indian farmers will no doubt (and with some justification) claim that their livelihoods have been unfairly affected by this change in policy. In India, as we have outlined, this will come up against the clear dictates of the constitution, which embeds sustainability as a national priority. Is this a unique problem for developing country farmers? Undoubtedly not, because the only uniqueness lies in the severity of the likely consequences of the reduction of previously available water resources.

In Australia, the issue of unsustainable water use and agriculture has attracted attention, particularly as the result of prolonged drought in much of the country. This provides a useful illustration of the common features the social licence and legal entitlement issues of water access in both India and Australia. Cubbie Station, Australia's largest cotton farm, near Dirranbandi on the border of New South Wales and Queensland, sits at the northern end of one of the country's most heavily debated environmental icons – the ailing Murray–Darling River system (Osborne 2009). Cubbie Station was founded in 1965, to dam the Culgoa River, store the equivalent of Sydney Harbour in open lakes, and irrigate up to 33 000 hectares (Sharrad 2009). The station's permits to divert and store more than 500 000 megalitres of water affect downstream river users in New South Wales, Victoria and South Australia.[23] With this water, Cubbie is able to grow about 13 000 hectares of irrigated cotton and generates about $50 million a year.[24] Investors, reliant on the legal rights to extract this water, invested substantial amounts to create this enterprise. However, with changing natural conditions, improved scientific knowledge and shifting community perceptions has come a new national water policy. Bringing water extraction back to within sustainable limits has become the fundamental goal of the Australian *Water Act 2007* (Cwlth).

Cubbie became a contentious issue in the national debate for water in Australia, with the farmers downstream accusing giant cotton farms of siphoning off a major share of water, which was historically flowing from Queensland into the NSW river systems. The pressure increased on the federal government to purchase the water licence so that enough water could be left in the Murray–Darling River for environmental flows.

In late 2009, Cubbie Station was put up for sale to reduce debt and made an eleventh hour deal offer to sell 25% of its vast water entitlements to the federal government, at a cost to taxpayers of $50 million (Bita 2010). For some, farmers this action was seen as a fundamental attack on farmers' water rights. For others, it was seen as a significant victory in defending these rights.

Conclusion

One of the goals of the sustainable agriculture movement is to create farming systems that mitigate or eliminate environmental harms associated with industrial agriculture. Sustainable agriculture is part of a larger movement towards sustainable development, which recognises that natural resources are finite, acknowledges limits on economic growth and encourages equity in resource allocation. Sustainable agriculture gives due consideration to long-term interests, such as preserving topsoil, biodiversity and rural communities, rather than only short-term interests, such as profit. Sustainable agriculture is holistic in that it takes a system-wide approach to solving farm management problems, and also because it places farming within a social context and within the context of the entire food system (Horrigan *et al.* 2002). This approach is closely aligned with the concept of the social licence to farm.

This chapter has looked at a variety of social licence issues in both developed and developing country contexts. Beneath the diversity of the examples, we have seen some important common features. The first is the prevalence of conflict between farming and other demands on natural resources. Whereas once it was almost automatic that food and fibre production would prevail as a resource use, this is now more heavily contested. The second is that the fundamentals of the political and institutional settings in the society are very important in determining how such contests will be played out. The third is that property rights, while tactically significant, are not the determinants of the outcome of such contests.

The examples also suggest that ultimately the outcomes for farmers depend less upon tactics that they use to advance their claims of a social licence and far more upon the objective performance of the sector in genuinely delivering the maximum of economic and social welfare to their host communities, with the absolute minimum of ecological and social harms. Where this equation is favourable, and the institutional settings strong, then farmers are more likely to see their ostensible private property rights supported by public institutions.

Acknowledgements

The authors of this chapter gratefully acknowledge the research assistance of their PhD student, Ms Hem Lata.

Endnotes

1 Environment and early patterns of adaptation, India: Earliest Times to 800 A.D., Indira Gandhi National Open University, 2005.
2 Quoted from 'Farmers suicides in Andhra Pradesh', AWARE Development Research Advisory Group, undated.
3 See, for example, Maneka Gandhi v Union of India and others and Bandhua Mukti Morcha v. Union of India and others.
4 See, for example, Unni Krishnan v. State of A.P.
5 See, for example, Subhash Kumar v. State of Bihar, Virendra Gaur v. State of Haryana, B.L. Wadhera v. Union of India and Vellore Citizens Welfare Forum v. Union of India etc.
6 Writ Petition (Civil) No.967/1989 (Indian Council for Enviro-Legal Action and Others Vs Union of India & Ors.).
7 Indian Council for Enviro-Legal Action and Others Vs Union of India & Ors. (1996) 3 SCC 212 p. 217.
8 Ibid.
9 Vellore Citizens Welfare Forum v. Union of India & Ors. (1996) 5 SCC 647 p. 657.
10 Ibid.

11 Ibid.
12 See <http://www.friendsoflafti.org/About_LAFTI.php>.
13 S. Jagannathan v. Union of India, WP 561/1994.
14 S. Jagannathan v. Union of India & Ors. (1997) 2 SCC 87 p. 139.
15 *The Constitution of the Kingdom of Bhutan* art 5.
16 Ibid.
17 *Bhutan: National Action Program to Combat Land Degradation* (2009), <http://www.undp.org.bt/assets/files/publication/NAP_Draft_Full&Final_Oct09.pdf> at 28 June 2010.
18 Ibid.
19 WWF, Thirsty crops, our food and clothes: eating up nature and wearing out the environment?, Living waters, conserving the source of life, <http://assets.panda.org/downloads/wwfbookletthirstycrops.pdf> at 28 June 2010.
20 Ibid.
21 Ibid.
22 Zhu T, Ringler C and Cai X (undated) 'Energy price and groundwater extraction for agriculture: exploring the energy water food nexus at the global and basin levels'.
23 Water sucking Cubbie Station for sale', *ABC News*, Aug 17, 2009.
24 The rise and rise of Cubbie Station, Melaleuca Media, <http://www.melaleucamedia.com.au/01_cms/details.asp?ID=257> at 28 June 2010.

References and further reading

Bita N (2010) 'Cubbie's 11th-hour, $50 m deal for Wong', *The Australian*, June 16.

Gupta AK (2004) Origin of agriculture and domestication of plants and animals linked to early Holocene climate amelioration. *Current Science* **87**(1), 54–59.

Horrigan L, Lawrence RS and Walker P (2002) How sustainable agriculture can address the environmental and human health harms of industrial agriculture. Environmental Health Perspectives **110**(5), 445–456.

Linthoingambi Devi N, Qi S, Chandra Yadav I, Dan Y and Fang T (2009) Pesticides in China and its sustainable use – review. Paper presented at Sustainable Land Use and Ecosystem Conservation, May 4–7, 2009, UNESCO, Beijing, P.R. China.

Meeta and Rajivlochan (2006) Farmers suicide: facts and possible policy interventions. Yashada, Pune.

Mehta MC (2009) *In the Public Interest*, Vol. 1 Prakriti Publications, New Delhi.

Osborne P (2009) Cotton brawl has many threads. *Brisbane Times*, August 21.

Sainath P (2009) The largest wave of suicides in history. *Counter Punch*, February 12.

Shan S (2010) Pesticide residues on vegetables from China. *The Epoch Times*, June 21.

Sharrad P (2009) The river is three-quarters empty: some literary takes on rivers and landscapes in India and Australia. University of Wollongong.

Sharma A and Banerji R (1996) The blind court. *Down to Earth* April 15–30.

Sustainet (2006) Sustainable Agriculture: A Pathway Out of Poverty for India's Rural Poor. <http://www.sustainet.org/download/sustainet_publication_india_part1.pdf> at 27 June 2010.

Yang Yang (2007) A China Environmental Health Project Factsheet, Pesticides and Environmental Health Trends in China. <http://www.wilsoncenter.org/topics/docs/pesticides_feb28.pdf> at 21 June 2010.

7

Retaining the social licence: the Australian cotton industry case study

Guy Roth

The Australian cotton industry has been no stranger to community and social anxiety. This was particularly evident throughout the 1980s and early 1990s, with outcry about the industry's pesticide usage and its impact on community health, water quality, fish kills and pesticide residues in beef cattle. Other aspects of cotton growing, such as fertiliser use, alteration to river flows, soil degradation and vegetation clearing, had also attracted the attention of the environmental regulatory agencies. In the late 1990s, the use of genetic modified traits in varieties kept cotton in the media headlines, while in more recent years during the ongoing drought public attention largely shifted to the industry's use of water for irrigation.

This chapter outlines proactive initiatives, and their outcomes, of the Australian cotton industry that enabled it to defend its *social licence to operate* when, in the early 1990s, community pressure went close to forcing the federal government to impose a regulatory regime that may have proven uneconomic for the industry.

Background to the Australian cotton industry

Cotton is the most commonly produced natural fibre in the world and represents just under half of the world textile market. On a global scale, Australia is a relatively small producer of cotton, growing about 3% of the world's cotton. The largest producers are currently China, India, USA, Pakistan, Brazil and Uzbekistan.

There are around 800 cotton farmers in Australia and approximately 10 000 people employed by the industry. Seventy per cent of Australia's cotton is typically grown in New South Wales, with the remainder grown in Queensland (Figure 7.1). Cotton is a major source of regional economic activity where it is grown, and usually generates 30–60% of the gross value of all regional agricultural income where it is produced, which makes up 10–30% of the gross regional product (Roth 2010). Its indirect impact on local economies is high.

Depending on water availability, up to 400 000 hectares of irrigated cotton can be grown in Australia. The area of rain-grown or dryland cotton changes considerably from year to year, depending on rain and cotton prices. The dryland area ranges from 5000–120 000 hectares, and this cotton is produced by up to 450 growers. Since 1980, the value of Australian cotton produced annually has increased dramatically to around $1.4–$1.6 billion per annum. In recent years, this gross value of production fell to less than $1 billion because of drought conditions. As in some of its competitor's market places, there is no direct government intervention or market support mechanisms applied to the growing or marketing of cotton in Australia. Australian cotton yields have increased significantly each year (Figure 7.2). During the last 20

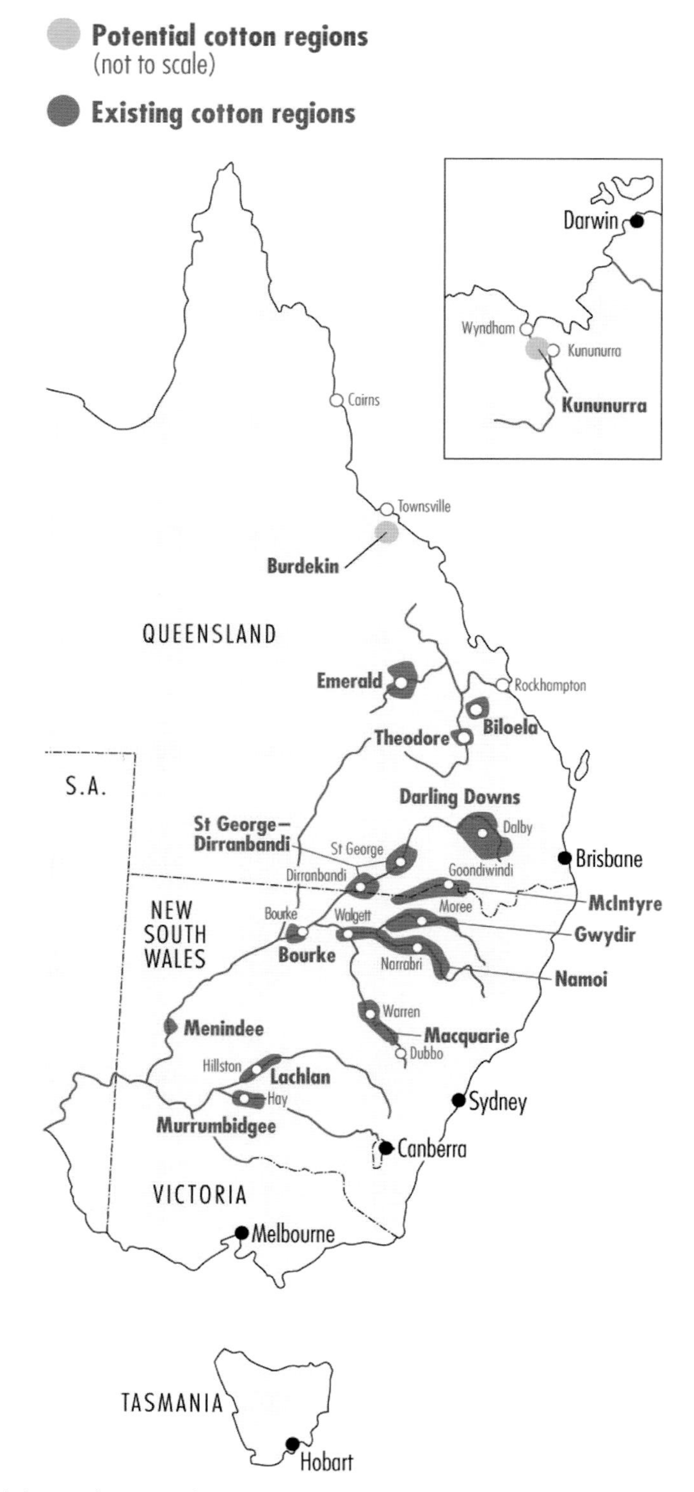

Figure 7.1: Existing and potential cotton regions in Australia (Source: Roth 2010)

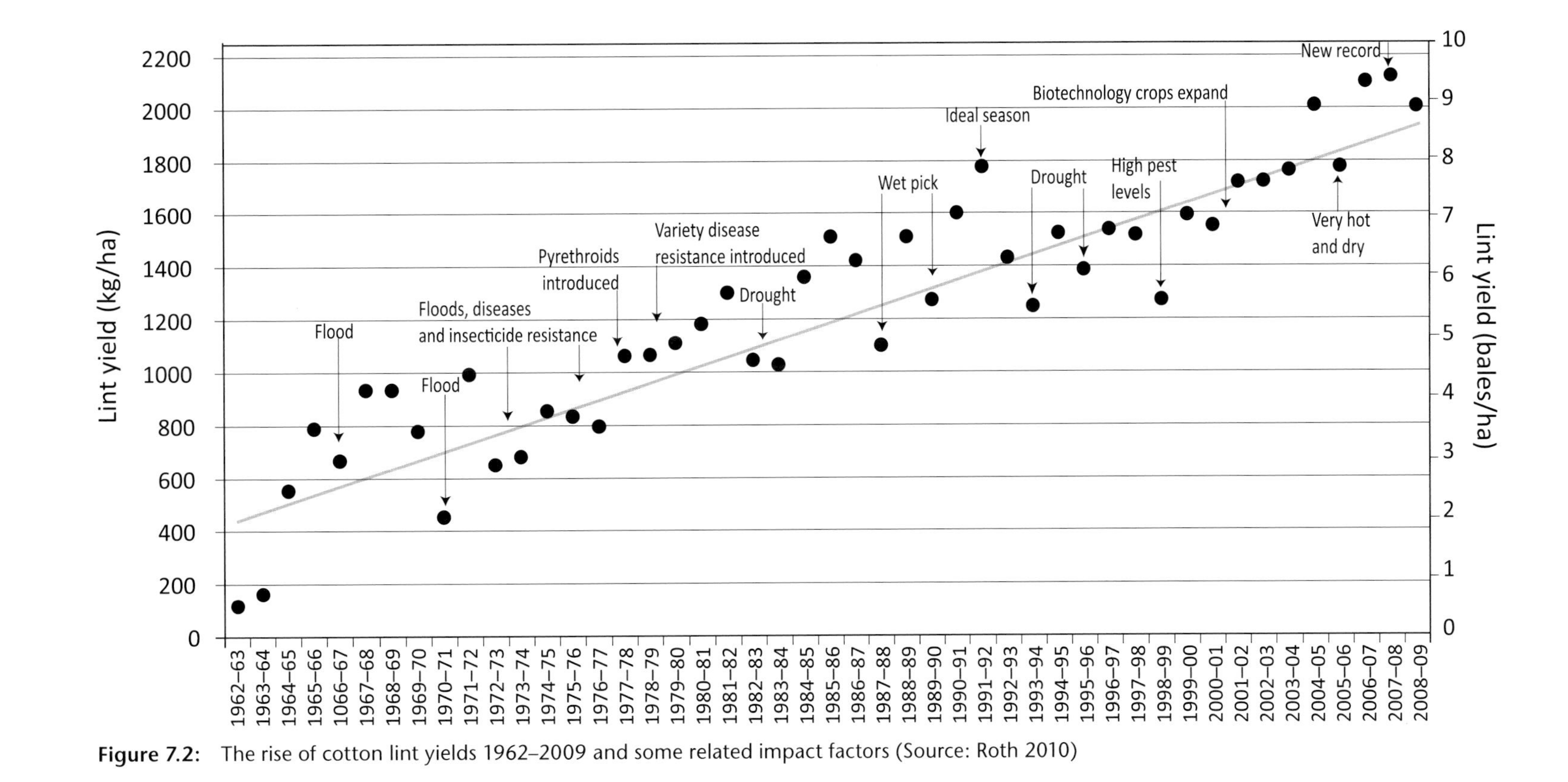

Figure 7.2: The rise of cotton lint yields 1962–2009 and some related impact factors (Source: Roth 2010)

years, cotton yields have increased on average at 32.9 kg of lint per hectare per year. Australian cotton yields are about three times greater than the world average (Roth 2010).

Australia has a reputation for producing high-quality cotton. This quality generally compares very favourably with that from other cotton-producing nations (Vijayshankar 2006; Dall'Abra 2006; Van der Sluijs *et al.* 2004; Shimazaki 2008; Yung 2008). The major buyers of Australian cotton are China, Indonesia, Thailand, South Korea and Japan. Australian cotton is increasingly considered a niche product because of its high quality in the world market, but the industry agrees that there are still opportunities for improvement. In this high-end market, Australia's major competitors are Texas and California in the USA and other longer staple producers such as China, India, West Africa, Uzbekistan and Brazil.

Customer demand is a key determinant of the type of cotton grown in Australia. An important trend in the world textile market is environmentally friendly cotton or 'eco cotton' (Fitzpatrick 2008; Spellson 2008; Yung 2008). Organic cotton is part of the 'eco cotton' theme, but cannot be produced in a cost-effective manner in large quantities. Hence, what has emerged is the importance of cotton produced according to environmental standards that have market traceability. Australia is one of the few countries where this can be delivered to market with complete reliability and transparency (Shimazaki 2008).

Responding to community concerns to regain the social licence

The cotton industry recognised it needed to be proactive in tackling environmental and community concerns, which had risen to a critical point in the early 1990s. It was considered important not to rush into a strong regulatory response because of concern of severe impact to the cotton industry economically (Schofield 1998). The cotton industry therefore initiated several key actions:

- It funded an external environmental audit of the industry in 1991. It commissioned an industry-wide environmental audit using independent international consultants Gibb Environmental Science and Arbour International. The audit report identified a suite of opportunities for improvement.
- It commissioned a scientific review on the effects of pesticides on the riverine environment (Barrett *et al.* 1991).
- It conducted a series of workshops with a range of stakeholders to identify research and development opportunities.

The urgent need for research and development solutions was led by the Cotton Research and Development Corporation (CRDC) who, in collaboration with the then Land and Water Resources Research and Development Corporation and the Murray–Darling Basin Commission, conducted a research program titled 'Minimising the impact of pesticides on the riverine environment using the cotton industry as a model'. This research program ran from 1993 to 1998, investigating the ways in which pesticides could move off the farm and the best ways of managing this movement, so that contamination risks could be minimised. A detailed description of the program can be found in Schofield *et al.* (2005).

Armed with more detailed knowledge, the cotton industry moved to identify solutions by holding a workshop. Somewhat surprisingly, 70% of the potential solutions were drawn from farmer knowledge (often comprising straightforward actions) while 30% came from new research (Schofield *et al.* 2005). These actions were then compiled into a formal Best Management Practices (BMP) Manual for managing pesticides. This BMP Manual has since been expanded to cover issues other than pesticides, such as water, soil and other natural resources.

The Cotton Best Management Practices Program

The cotton industry's BMP Program is a voluntary environmental risk management program based on a process of continuous improvement using a 'plan–do–check–review' management cycle. Cotton Australia (2006) describes it as a 'functional environmental management system'. An overview of the Cotton BMP program by Cotton Australia (2006) outlined the program goals as:

> *'To see the development of the cotton industry:*
> - *whose participants are committed to improving farm management practices*
> - *whose participants have developed and follow policies and farm management plans that minimise the risk of any adverse impacts on the environment or human health*
> - *which can credibly demonstrate to the community stewardship in the management of natural resources and farming operations.'*

The Cotton BMP Program was introduced to primarily improve the management of pesticides (Williams and Williams 2001; Williams *et al.* 2004; Ross and Galligan 2005; Schofield *et al.* 2005). The BMP program then evolved to consider broader natural resource management issues related to land and water management. Now, with the launch of the revised program in 2010, it seeks to provide a whole-of-farm tool for growers to manage their enterprises and the everyday risks of farming. The Cotton BMP Program presents the opportunity for the cotton industry to provide more confidence to the community, governments and cotton markets in its ability to use and manage various technologies such as pesticides and gene technology (Anthony 2004).

The BMP Program also provides a systematic process for the cotton industry and its growers to contribute to the catchment planning and natural resource management goals of government. It is a proactive initiative that is enhancing cotton growers' social licence to farm (Higgins and Adcock 2008).

In 2005, the industry embarked on the legal process to acknowledge the social licence afforded to industry by BMP. The industry embarked on the unique pathway of attempting to have the BMP Program formally recognised by the Queensland Government as delivering outcomes consistent with its statutory Land and Water Management Plans. These have the purpose of 'providing individual landholders with a practical management plan that demonstrate that water use practices are ecologically sustainable, both on and off farm' (Department of Natural Resources and Mines 2001). This achievement was realised in December 2008, with the Cotton BMP Program currently being the only voluntary on-farm management system recognised by a government in Australia as an alternative pathway for landholders to achieve a regulatory requirement. The importance of a social licence in the industry was reiterated in 2008, by the Chairman of the Australian Cotton Industry Council BMP Committee who stated that the program was a proactive initiative to help maintain its social licence to farm (Cotton Australia 2008).

Most recently, the cotton industry has been investigating the application of BMP in the post-farm-gate sectors of ginning and transport. The implementation of BMP at the grower level and the use of it throughout the post-farm-gate supply chain provides a vehicle and standards for the improvement of Australian cotton product (Dall'Alba 2006). The cotton industry has investigated the cotton market requirements of retailers (Williams 2007), with work continuing by the Australian Cotton Shippers Association and Cotton Australia to evaluate the promotion and marketing of 'BMP cotton' as environmentally responsible cotton

(Spellson 2008) and capitalise on this global focus of environmentally sound production of food and fibre.

Monitoring of Cotton BMP outcomes

Since the introduction of the Cotton BMP Program in 1997, independent reviews have found that at least 85% of cotton growers have changed their practices as a result. In May 2006, Cotton Australia undertook a survey of 70 growers, including levy and non-levy payers, BMP participants and non-BMP participants (Cotton Australia 2006). The result showed that 79% of the growers felt that BMP had improved the environmental performance on their farm, 31% of growers felt BMP had improved the financial performance of their farm, and 46% indicated it had improved staff management.

Cotton Australia also asked the growers what they thought the industry could do to get other growers to adopt BMP. Growers identified the key ways for industry to support growers in adoption of BMP is to demonstrate its benefits and develop grower champions and grower-to-grower encouragement, as well as providing incentives and discounts.

The voluntary audit process has posed challenges in its management (Hassall and Associates 2006), implementation and adoption. Auditing is often stated as a barrier and unnecessary aspect of the BMP program. However, a review by Holloway and Roth (2003) of grower feedback on audits found that 90% of respondents felt an audit was of significant benefit. Some grower comments on the benefits of the audit program included: '... it makes you aware of your obligations, it focused on the issues we overlooked, it gave us the push to do things we have been putting off.'

In 2003, CRDC commissioned GHD Pty Ltd to conduct the second environmental audit of the Australian cotton industry (GHD 2003) and to assess the industry's response to the previous environmental audit in 1991 (Gibb 1991). The second environmental audit involved a review of the literature, workshops with stakeholders and visits to 32 farms. It noted significant improvements in farm management practices.

As noted in the second *Environmental Audit of the Cotton Industry:*

> *'One of the most significant environmental improvements in the Australian Cotton Industry was the development of the BMP program. The audit identified a direct link between areas of improvement observed on the properties and the BMP modules. Farms that had undertaken their second BMP audit showed real improvements in environmental management and the auditing process provided a benchmark to indicate that progress had been made. The BMP audits were found to give a good assessment of the environmental farm practices currently covered by the manual.'*

In 2003, Macarthur Agribusiness was commissioned by CRDC and Cotton Australia to undertake an evaluation of BMP outcomes. The evaluation involved 10 farm visits, 65 telephone interviews and focus groups in five cotton regions. The findings of the report are as follows:

- Significant beneficial change in cotton farm practices had taken place since the manual was introduced in 1997, such as improvements to IPM, pesticide application, communication, weather monitoring, reduced pesticide use, reduced spray drift and odour complaints, improved water quality and a reduction in fish kills and cattle contamination.
- On-farm economic outcomes were difficult to quantify. This finding was similar to a conclusion reported by Cotton Australia (2006). Actions undertaken were often viewed as things growers would have done anyway.

- External stakeholders regarded the audit program as important.
- It was recommended that audit data be used for triple bottom line reporting.

Hassall and Associates (2006) evaluated the implementation of the BMP process and in particular the BMP Land and Water Module. The study found that growers and stakeholders considered the BMP process and the Land and Water Module to be effective, with well-developed tools for reviewing and planning changes to activities on farm. It also found the Land and Water Module effectively deals with most key natural resource management issues relevant to the cotton industry and made several recommendations to improve BMP uptake by growers. Likely outcomes include changes in attitude, knowledge and aspirations, as well as natural resource management outcomes such as water use efficiency and soil health. Protection of the right to farm and continued access to water were found to be the largest potential benefits for production and profitability.

An example of a cotton grower quote from Hassall's report (2006) highlights the feeling that growers understand it is important to demonstrate externally that they are being responsible, which in turn helps to support a social licence to operate: 'BMP is important so that cotton can demonstrate that they're doing things well as an industry'.

A farm agronomy adviser noted that BMP was not about reducing costs or increasing production, but rather about being able to continue farming: 'We need to keep growers in business – BMP helps to do that'.

Roth (2010) completed an analysis of the Cotton BMP program farm practice audit criteria for the 10 years between 1999 and 2008. Results show it is possible to quantify how cotton growers have implemented changes to a wide range of their farm management practices. The analysis showed there was a very high standard of legal compliance on farms during 1999–2008 where the BMP program was adopted. Figure 7.3 shows the mean BMP ranking for all 47 farm practice criteria from the pesticide application, pesticide storage, integrated pest management, farm design and farm hygiene modules for the 10 years between 1999 and 2008. The rankings averaged 1.46 (scale 1–4) and showed a 29% improvement over the decade. It showed a 45% improvement between 1999 and 2006. The fall in the mean BMP farm practice standards between 2007 and 2008 was attributed to the ongoing drought, which reduced expenditure, action and motivation.

Despite the drought, the BMP farm practice standards for the five years (2004–2008) were on average better than the previous five years (1999–2003). The analysis showed the mean BMP ranking for certified audited farms between 2006 and 2008 was 24% better than the non-audited farms. This supports the premise that the extra rigour associated with external audit does lead to additional on farm improvements in practice.

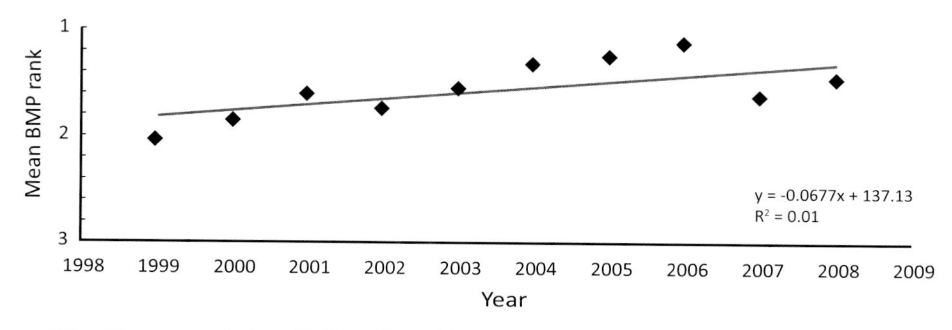

Figure 7.3: The mean Cotton BMP audit rankings for all 47 farm practice criteria between 1999 and 2008 from the pesticide application, pesticide storage, integrated pest management, farm design and farm hygiene modules (Source: Roth 2010)

Figure 7.4 shows significant improvements were made between 1999 and 2008 in pesticide spill containment (69%), storage ventilation (49%), security (51%), work procedures (19%) and emergency procedures (39%). Figure 7.5 shows significant improvements were made between 1999 and 2008 related to pesticides in signage on chemical storages (38%), mixing and loading sites (62%), mixing and loading systems (65%), worker safety (22%) and waste disposal (35%). Trends for equipment maintenance and safe transport are not clear, but were of a high standard.

Reduced insecticide use

Endosulfan is an organochlorine insecticide used to control sucking, chewing, and boring insects and mites in a range of crops, including cotton and sorghum. Figure 7.6 illustrates that endosulfan concentrations, which were very high in 1991, have been below the Australian and New Zealand water quality guideline trigger value for 99% ecosystem protection (ANZECC and ARMCANZ 2000) for the last seven years (Mawhinney 2008). The adoption of the Cotton BMP Program improved tail water return systems. Restrictions placed on endosulfan use and the introduction of genetically modified 'Bt' cotton has all contributing to the reduced movement of endosulfan into river systems. Similar results have been reported in the Gwydir and Macintyre Valleys during 1992–2003 (Mahwinney 2004) and in the Queensland Murray–Darling Basin during 1994–2001 (Waters 2004).

Number of complaints to the EPA

Complaints to the Environment Protection Authorities (EPA) are a good indicator of the absence of strong community dissatisfaction in relation to industry performance. There has been a dramatic drop in the number of complaints to the NSW EPA since 2001, which were down to three per year for 2006 and 2007. This can been attributed to a number of linked factors including the implementation of the Cotton BMP program, greater use of transgenic cotton varieties and a reduction in the crop area due to the drought. Fewer complaints lead to greater social harmony in the community, which in turn leads to less threats to the farmers' social licence to operate.

Community attitudes and the social licence

Community attitudes are important because they influence the social licence to farm. Between 1995 and 2004, Cotton Australia commissioned five studies that investigated community attitudes towards the Australian cotton industry. The studies were carried out by professional attitudinal research companies, namely Stollznow Research and Roy Morgan Research (Stollznow 1995a, 1995b, 1997, 1998; Roy Morgan Research 2004). The issues raised in these studies included:

- community health: harmful chemicals, chemical smells, aircraft noise and spraying, beef cattle contamination by Helix and endosulfan, and soil contamination
- pesticides, herbicides and defoliants: excessive use and spray drift
- river water: chemicals in the water and run off, high water use and salinity
- groundwater: excessive use drying up stock bores and entry of chemicals
- soil: exploiting soils, chemicals and residues
- land clearing and laser levelling
- cotton growers: perception of being greedy, arrogant, irresponsible and only in it for the short term
- cotton industry: perceived as all powerful, secretive and dishonest
- the Cotton industry was rated consistently low in surveyed attributes compared with other industries

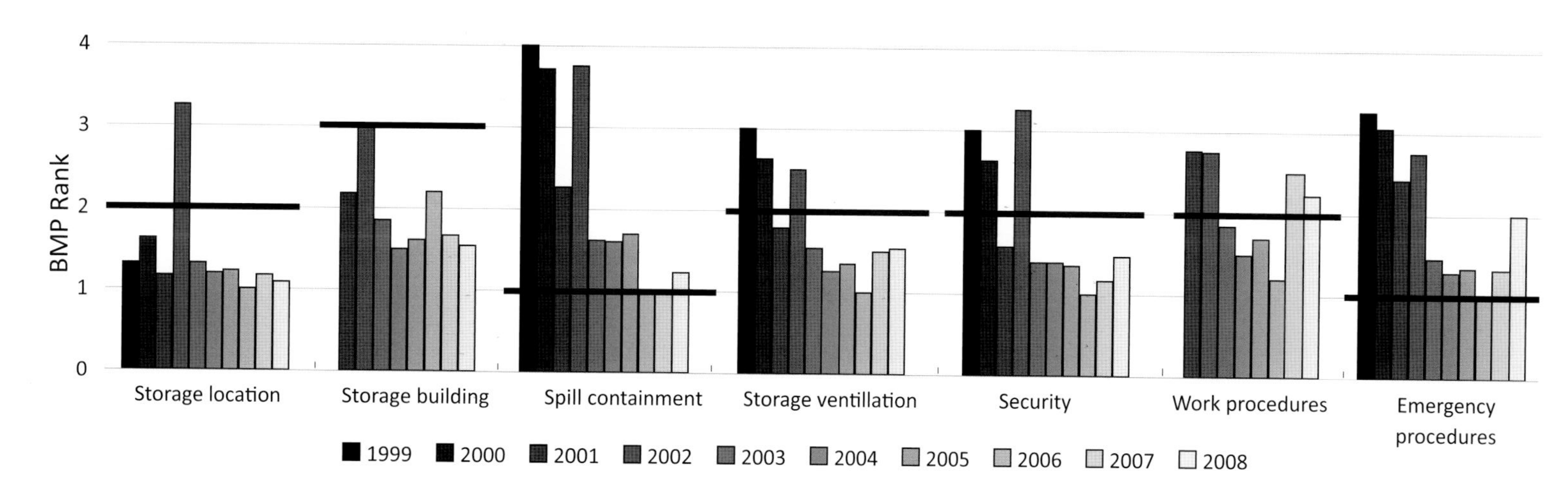

Figure 7.4: The mean BMP rankings for storage location, storage building, spill containment, storage ventilation, security, work procedures and emergency procedures on BMP audited cotton farms over 10 years between 1999 and 2008 (Source: Roth 2010). The horizontal bar is the BMP compliance standard. Rank 1 is excellent. Rank 4 is poor.

Figure 7.5: The mean BMP rankings for signage, mixing and loading sites and systems, worker safety, equipment maintenance, waste disposal and safe transport on BMP audited cotton farms over 10 years between 1999 and 2008 (Source: Roth 2010). The horizontal bar is the BMP compliance standard. Rank 1 is excellent. Rank 4 is poor.

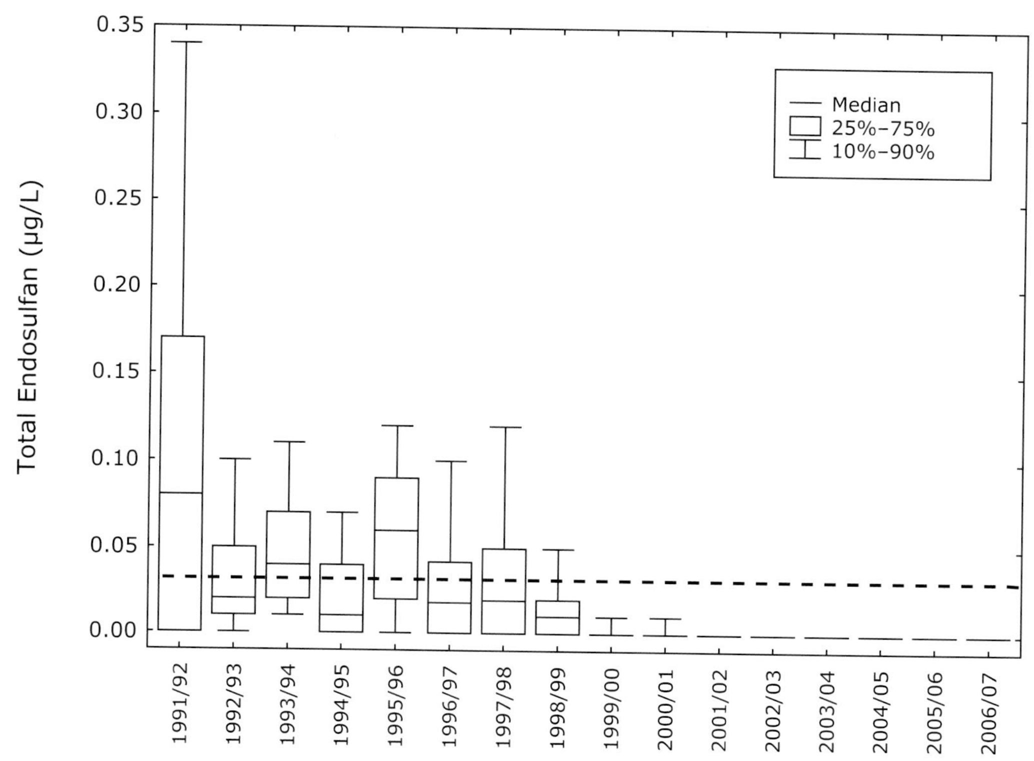

Figure 7.6: Total endosulfan concentrations in the Namoi Catchment from 1991–1992 to 2006–2007 (Source: Mawhinney 2008). The broken line represents the Australian and New Zealand water quality guideline trigger value (ANZECC and ARMCANZ 2000) for 99% ecosystem protection (0.03 µg/L). Each box represents the middle 50% of the data collected for each year. The middle line in each box represents the median (or 50th percentile) value, which is the most useful when assessing water quality.

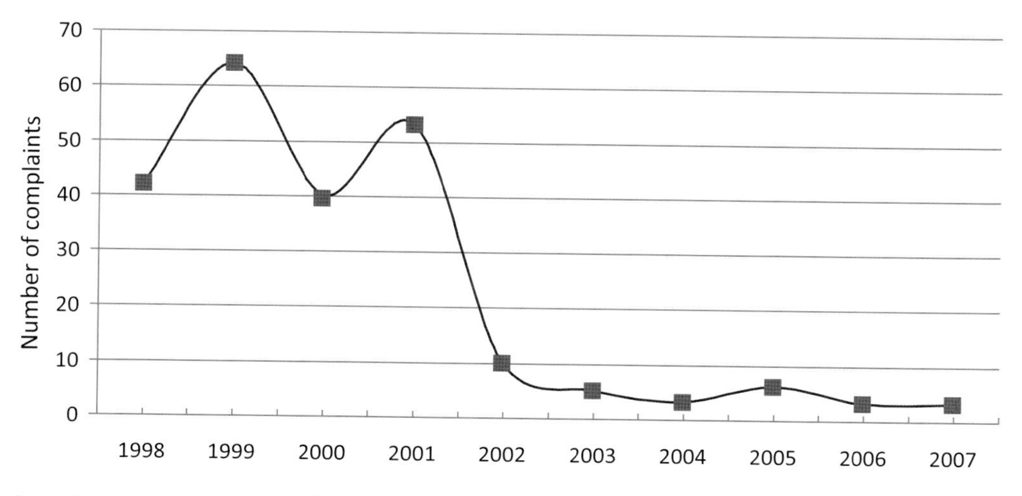

Figure 7.7: Number of complaints received by the NSW EPA 1998–2007 (Source: NSW EPA *pers. comm.* 2007)

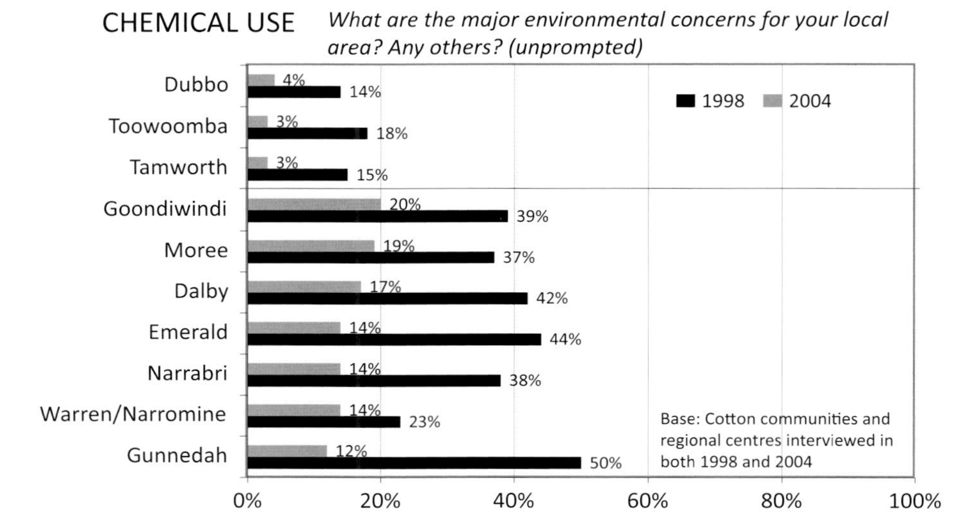

Figure 7.8: A comparison of community concerns in cotton growing regions regarding agricultural chemical use between 1998 and 2004 (Source: Roy Morgan Research 2004)

- Moree and Gunnedah were noted as towns where there was most community negative orientation towards the cotton industry.

In 2004, Cotton Australia and CRDC commissioned further attitudinal research into the cotton industry by Roy Morgan Research (2004). This study included major cotton towns, a number of large regional centres nearby the cotton communities, but themselves not cotton towns (Dubbo, Toowoomba and Tamworth) . Community member's responses to the cotton industry in both 1998 and 2004 were reported. In 1998, chemical use was still a major concern, but by 2004 this had reduced significantly in all centres (Figure 7.8). This research also showed that community concerns about the cotton industry's chemical use, spray drift and water use had reduced significantly between 1998 and 2004.

It should be noted that these changes in community perceptions, do, in principle, mirror actual improvements in cotton farming practices that have been demonstrated through on-farm audit and farmer surveys.

Looking to the future

The Australian cotton industry, like any other industry, will need to balance economic, environmental and social sustainability issues. The cotton industry has a number of initiatives underway aimed at better managing these competing pressures to improve its public standing and maintain its social licence to operate. Research and development and the systematic implementation of better practices by farmers via the BMP Program are the central planks of the strategy. Marketing or spin campaigns are not part of the strategy.

The cotton industry has produced a vision strategy out to the year 2029 (CRDC 2010), which was launched at the 2010 Australian Cotton Conference. A key element of this vision is being a responsible producer and supplier of the most environmentally and socially responsible cotton in the globe. Other key elements of the 'Preferred Future for 2029' envisage an industry that represents 'Australian cotton, carefully grown, naturally world's best': an industry that is differentiated, responsible, tough, successful, respected and capable. To achieve this, the industry needs good measures of its economic, environmental and social performance for the community to deliberate.

References and further reading

Anthony D (2004) Research and development opportunities. *Proceedings of 12th Australian Cotton Conference,* 10–12 August 2004, Broadbeach, Queensland. pp. 39–56.

ANZECC (Australian and New Zealand Environment and Conservation Council) and ARMCANZ (Agriculture and Resource Management Council of Australia and New Zealand) (2000) 'Australian water quality guidelines for fresh and marine waters, national water quality management strategy'. ANZECC, Canberra.

Barrett JWH, Peterson SM and Batley GE (1991) 'The impact of pesticides on the riverine environment with specific reference to cotton growing'. Cotton Research and Development Corporation and Land and Water Resources Development Corporation, Canberra.

Cotton Australia (2006) 'Background information for the Cotton Industry BMP review meeting 13th June 2006'. Unpublished report, Cotton Australia, Sydney.

Cotton Australia (2008) 'Strategic plan for BMP cotton'. Cotton Australia, Sydney.

CRDC (Cotton Research and Development Corporation) (1997) *The Cotton Industry Best Management Practices Manual.* 1st edn. Cotton Research and Development Corporation, Narrabri, NSW.

CRDC (2000) *The Cotton Industry Best Management Practices Manual.* 2nd edn. Cotton Research and Development Corporation, Narrabri, NSW.

CRDC (2010) 'Vision 2029'. Cotton Research and Development Corporation, Narrabri, NSW.

Dall'Alba B (2006) Marketing our product. *13th Australian Cotton Conference,* 8–10 August 2006, Broadbeach, Queensland. pp. 41–50.

Department of Natural Resources and Mines (2001) 'Guidelines for land and water management plans'. Queensland Government, Brisbane.

Fitzpatrick C (2008) What do consumers want from textiles? *Proceedings 14th Australian Cotton Conference,* 12–14 August 2008, Broadbeach, Queensland.

GHD (2003) 'Second Australian cotton industry environmental audit: a report for the Cotton Research and Development Corporation'. Cotton Research and Development Corporation, Narrabri, NSW.

Gibb (1991) 'An environmental audit of the Australian cotton industry'. Gibb Environmental Sciences and Arbour International.

Hassall and Associates (2006) 'Evaluation of the Australian Cotton Industry's BMP process'. Unpublished report prepared for Cotton Research and Development Corporation and Cotton Australia, Narrabri, NSW.

Higgins S and Adcock L (2008) New beginnings for BMP. *Proceedings 14th Australian Cotton Conference,* 12–14 August 2008, Broadbeach, Queensland.

Holloway R and Roth G (2003) Grower feedback on Cotton BMP auditing. *The Australian Cotton Grower Magazine* February–March, 20–22.

Mahwinney W (2004) Water quality in the Namoi. In *WATERpak: A Guide for Irrigated Management in Cotton and Grains.* pp. 265–272. Cotton Research and Development Corporation and Cotton Catchment Communities CRC, Narrabri, NSW.

Mahwinney W (2008) 'Water quality in the Barwon region'. Department of Water and Energy, Tamworth, NSW.

Ross C and Galligan D (2005) The integration of Cotton BMP with a sub catchment planning approach. *4th National EMS in Agriculture Conference*, 17–20 October, Beechworth, Victoria.

Roy Morgan Research (2004) 'Attitudinal research in cotton communities and opinion centres'. A report for Cotton Australia and Cotton Research and Development Corporation. Cotton Australia, Sydney.

Roth GW (2010) 'Economic, environmental and social indicators of the Australian cotton industry'. Occasional Report, Cotton Catchment Communities Cooperative Research Centre and Cotton Research and Development Corporation, Narrabri, NSW.

Schofield N (1988) Origins and design of the cotton pesticides program. In: *Minimising the impact of pesticides on the riverine environment: key findings from research with the cotton industry*. Occasional paper 23/98, Land and Water Resources Development Corporation, Canberra.

Schofield N, Williams A, Holloway R and Pyke B (2005) Minimising riverine impacts of endosulfan used in cotton farming – a science into practice success story. *Proceedings Pacifichem Symposium 2005*, Dec 15–20 2005, Honolulu, USA.

Shimazaki T (2008) Winning back markets for Australian cotton. *Proceedings of 14th Australian Cotton Conference*, 12–14 August 2008, Broadbeach, Queensland.

Spellson A (2008) What we are doing with your BMP Cotton. *Proceedings 14th Australian Cotton Conference*, 12–14 August 2008, Broadbeach, Queensland.

Stollznow Research (1995a) 'Investigation and monitoring of consumer attitudes towards the cotton industry'. Report for the Australian Cotton Foundation (now Cotton Australia) Sydney.

Stollznow Research (1995b) 'Sydney community perception of the cotton industry and redated matters'. Report for the Australian Cotton Foundation, Sydney.

Stollznow Research (1997) 'Investigation and monitoring of community attitudes toward the cotton industry in country areas and qualitative investigation of community attitudes towards cotton growing'. Report for the Australian Cotton Foundation, Sydney.

Stollznow Research (1998) 'Investigation and monitoring of community attitudes toward the cotton industry in country areas and qualitative investigation of community attitudes towards cotton growing in NSW and Queensland towns'. Report for the Australian Cotton Foundation, Sydney.

Van der Sluijs MHJ, Gordon SG and Naylor GR (2004) Australian fibre quality: smart decisions. *Proceedings of 12th Australian Cotton Conference*, 1–12 August 2004, Broadbeach, Queensland.

Visijayshankar MN (2006) Processing your product using Australian cotton. *Proceedings of 13th Australian Cotton Conference*, 8–10 August 2006, Broadbeach, Queensland.

Waters D (2004) Water quality in Queensland catchments and the cotton industry. In: *WATERpak: A Guide for Irrigated Management in Cotton and Grains*. Cotton Research and Development Corporation and Cotton Catchment Communities CRC, Narrabri, NSW.

Williams A (2007) 'Enhancing the cotton industry's BMP Program to improve adoption'. Final report for EMS pathways program project, Cotton Research and Development Corporation, Narrabri and Department of Agriculture Fisheries and Forestry, Canberra.

Williams A, Leutton R, Rouse A and Cairns R (2004) The Australian cotton industry: turning natural resource management policy into on ground action. *OECD Expert Meeting on Farm Management Indicators and the Environment*, Palmerston North, New Zealand, 8–12 March 2004.

Williams A and Williams J (2001) 'Fostering best management practices in natural resource management – towards an environmental management system in the cotton industry'. Cotton Research and Development Corporation, Murray Darling Basin Commission, Canberra and Australian Cotton Growers Research Association, Wee Waa.

Yung L (2008) The changing world textile market. *Proceedings of 14th Australian Cotton Conference*, 12–14 August 2008, Broadbeach, Queensland.

8

Farmers heal the land: a social licence for agriculture in Iceland

Andres Arnalds

This chapter highlights common aspects of the issues affecting farmers' social licence to farm, contrasting the experiences of farmers and government in hot harsh climates of Australia and the colder climate of Iceland. These two countries are virtually polar opposites, with Iceland adjacent to the North Pole and Australia the closest continent to the South, with one country being characteristically wet and cool, and the other predominantly hot and dry. Yet, as this chapter illustrates, there are parallels and lessons that can be drawn by farmers in both countries that suggest a commonality of issues and approaches.

In both jurisdictions, farmers have seen a marked shift in community attitudes towards farming as a legitimate land use, changing from a purely economic and production emphasis to a growing expectation that farmers will demonstrate a land stewardship commitment. In both countries, the response has been translated through political processes into new legal and social requirements and a change in economic incentives to more deeply embed sustainability. In Iceland, as in Australia, conservation of soil and vegetation has emerged as a fundamental priority, and the use of native species has become a higher priority. One interesting difference is that the warming of the climate has, in Iceland, increased the farmers' productive capacity.

In Iceland, as in Australia, there has been a marked difference in the willingness and capacity of farmers to respond to changing social expectations. With this mixed adaptive capacity has come a varied ability to take advantage of economic opportunities that have come with changes in community expectations. In both countries, the most effective strategies have been those led by farmers in partnership with government and the community, notably through Landcare.

The lessons from this chapter are that the maintenance of the social licence to farm is important to primary producers, but achieving this is not simply a matter of following government dictates. Community attitudes are constantly changing, and the dynamics of politics, in turn, constantly accelerating. Keeping a finger on the pulse of society, accepting the need for change in how resources are used and conserved, and having an industry that can respond with positive leadership, are all essential elements affecting the ability of farmers to protect their social licence while remaining viable. This is the case regardless of whether one is running Merino sheep on the dry hot Hay Plains of NSW, or running Icelandic horses on the pastures beneath the cold Eyjafjallajokull, recently showered by the ash of an erupting volcano.

Social licence, regulation, trust and common resources are all linked. This has been no more convincingly demonstrated to us than with the meltdown of Iceland's banks, and the economic and social pain that this has created. There are lessons about social licence and farming that can also be drawn from this sorry tale.

Until the banking collapse in 2008, Iceland was one of the wealthiest countries in the world on a per capita basis. In that year, its private banks suffered massive losses during the early stages of the global financial crisis, leading to calls on Iceland's sovereign guarantees that were well beyond its public means to meet (for a detailed discussion of causes, see Jonsson 2009). The relationship between private entrepreneurial freedom to exploit a common resource (in this instance, our national credit) and the duty to be a good steward is, as a result, a matter of more than passing interest to Icelanders.

Iceland has ample experience of the tragedy of the commons, whether considering the land, sea or national finances. In the old days, the number of sheep was a measure of wealth, and in Icelandic the word 'fé' means both sheep and money. Historically, over-exploitation led to the collapse of the once fertile commons in the highlands. More recently, the defence of the cod fisheries, the ocean commons, against over-fishing led to stand-offs between tiny Icelandic Coast Guard vessels and British frigates in the 1950s and 1970s, known as the 'cod wars'.

Land condition and farming in Iceland

Iceland is the eighteenth largest island in the world, with a landmass of 103 000 km^2 and a population of around 330 000. East of the southern tip of Greenland and touching the Arctic Circle, Iceland has breathtakingly beautiful landscapes of mountains, volcanoes, rangelands and pastures, deserts and glaciers. The strong influence of the Gulf Stream results in a maritime cold temperate climate in the lowlands, to sub-arctic in the highlands. The weather is unstable. Strong winds and high rainfall events are frequent. Soil formation is greatly affected by a steady influx of aeolian materials from unstable desert surfaces and ash from frequent volcanic eruptions. These characteristics make the soils susceptible to erosion by wind and water (Arnalds *et al.* 2001).

The first humans who settled in Iceland, around 874 AD, were met by lush vegetation and fertile ecosystems. Up to two-thirds of the country may have been vegetated and at least 25% of the area was covered with woodlands, mainly birch (*Betula pubescens*). The first centuries of human history in the country was a time of prosperity, based largely on the fertility of an untouched country. However, population density soon increased and large-scale destruction of the natural resources began. Loss of woodlands, and the consequential soil erosion and other forms of severe land degradation, has reshaped the appearance of the Icelandic landscape. Most of the damage to Icelandic ecosystems is a consequence of a millennium-long interaction between unsustainable land use and natural forces in a sensitive environment.

The woodlands were cut for fuel and timber, or burned to provide space for agriculture and grazing. As their distribution and cover reduced, and grazing pressure increased, the sensitive volcanic soils lost their shelter and became more vulnerable to the forces of nature. Coupled with climatic fluctuations and volcanic eruptions, the pressure on the land exceeded its resilience, resulting in the ecosystem deterioration that Iceland faces today (Arnalds 1987).

The consequences of over-exploitation of the land for 1100 years are costly, both for current and future generations of Iceland. About 95% of the original woodlands and half of the vegetative cover may have been lost. Much of the remaining vegetation is severely degraded, biological diversity has been reduced, land fertility diminished and hydrology altered. Thorsteinsson (1986) estimates that the carrying capacity for livestock grazing is now only about 20% of what it may have been at the time of settlement.

At the time of human settlement about 1140 years ago, birch forest and woodland covered 25–40% of Iceland's land area. The relatively tall (to 15 m) birch forests of sheltered valleys graded to woodlands of birch and willow scrub towards the coast, on exposed sites and wetland areas, and to willow tundra at higher elevations (Eysteinsson 2009). As in agrarian societies everywhere, the settlers began by cutting down the forests and burning scrubland to create fields and grazing land. Sheep were important in Iceland from the outset, and sheep grazing had a direct effect on the birch woods and prevented regeneration. As a result, the area of woodland declined steadily, to a post-glacial minimum of less than 1% cover around 1950.

Most of contemporary Iceland can be classified as rangelands and pastures. Agriculture is by far the largest land user, and sheep and cattle for meat and milk accounts for 69% of the total value of agricultural produce. Production of sheep, cattle and horses are primarily based on grazing native pastures and rangelands, and making hay to feed the livestock through the long winters. Crop production is small scale, although rapidly increasing with new cultivars, especially barley, with the more favourable temperatures. The lowlands are divided into farms in private ownership, while the highlands are mostly divided into commons used for sheep grazing by the respective communities. Land condition varies, but in areas of severe land degradation grazing can have a dramatic effect. The result has been large-scale desertification in a humid environment. In degraded or denuded areas, livestock production can significantly slow vegetation recovery. Improvements in grazing management and revegetation efforts are needed in order to reach goals of sustainable land use in many areas of Iceland.

For 1000 years, Iceland was predominantly a subsistence agricultural society. In 1900, nearly 80% of the population still lived in rural areas. With the enhanced fish catch and developments in manufacturing and services, people increasingly moved from the country to towns and villages. By 2009, only 6.5% of the population lived in rural areas (Agricultural Statistics 2009). The workforce distribution between the employment sectors shows agriculture has fallen in proportion from 32% in 1940 to less than 3%. At the same time, the calculated share of agriculture in gross national product has fallen to a level of 1.4%. Despite its dwindling proportion in both these respects, agriculture still is of major significance in Iceland. Agriculture provides almost all milk and meat sold and a high proportion of other foods, such as vegetables. It is also culturally significant to many Icelanders, including urban dwellers.

Successes and failures in caring for the environmental commons

The massive land degradation and soil erosion that has taken place in Iceland demonstrates clearly how a vicious cycle of unsustainable land use, reduced land quality and struggle for survival can feed the loop of desertification in a self-sustainable farm-based culture. The main underlying causes are the same as in many other parts of the world – clearing of land by burning, over-harvesting of trees and scrubs and overgrazing – that together gradually weakened the resilience of ecosystems and hampered regeneration after disturbances.

The peak of the ecosystem destruction may have been reached in the 19th century, when new export markets for wool and wethers led to an increase in sheep numbers without regard for the carrying capacity of the country. Repeated disastrous events occurred in the last decades of the 19th century and numerous farms were destroyed by sandstorms and soil erosion (Olgeirsson 2007). Vast areas became denuded with a total loss of vegetation and soils. These catastrophic events triggered preventative measures, but ignorance of the role of grazing in this destruction and lack of means for erosion control prevented successful action.

With partial independence from Denmark in 1904, the tide started turning, with increasing realisation of the social costs of continued land degradation and its threat to the future of agriculture in many districts. In response to the immediate need, the *Act on Forestry and Protection against Soil Erosion* was passed by the Icelandic Parliament in 1907. This marked the beginning of organised battle against the rampant loss of vegetation and soil erosion that was destroying land quality across the country. By later amendments of this law, two state institutes were established: the Soil Conservation Service (SCS) and the Forest Service (FS) of Iceland.

Iceland's more than 100 years of national effort on halting soil erosion and restoring land quality are characterised by numerous success stories, despite limited financial and human resources for most of this time (Olgeirsson 2007; Arnalds and Runolfsson 2009). However, the task of protecting and restoring Iceland´s ecosystems is still enormous. Only a small proportion of the affected areas has been treated and serious soil erosion characterises about 40% of the land area of the country (Arnalds *et al.* 2001).

Overgrazing continues to be a major problem over extensive areas. Vegetation is still being lost through erosion. In addition to limited financial resources, Arnalds (2005b) argues that the reasons for the inadequate achievements on a national scale until after 1990 included governmental subsidies to sheep production that were lacking environmental safeguards. High level of government support led to an all-time peak in sheep numbers in the late 1970s, but poor grazing management resulted in severe overgrazing in many areas. As a result, the government was paying both for the damage to the land, and for its reparation. Weak laws on soil conservation prevented meaningful protection of sensitive soils and vegetation. The legal procedure for limiting grazing pressure was (and still is) too complex, rendering this option for preventive measures useless.

The mixed success of the early soil conservation work was influenced by many other factors, such as sociological barriers to improved conservation. A general lack of incentives to care for the land, and similarly a lack of disincentives to reduce unsustainable use, prevailed until fairly recently (Arnalds 2005b). Similar experiences have been described in many other countries. The top-down approach, lack of local involvement and focusing on symptoms rather than causes are among the organisational and strategic mistakes that were frequently seen, both in Iceland and in many other, widely differing countries (Roberts 1989; Douglas 1996; Hannam 2000; Sanders 2000).

Social licence as a driving force in farming

Agriculture in Iceland is significantly dependent on direct and indirect support from the government. Furthermore, a positive image of the agricultural environment and the quality of the production is crucial in maintaining healthy domestic markets and creating export markets in fierce competition with cheaper products.

Among OECD members, Iceland is in the group of countries with the highest national support for agriculture, although gradually reducing from 77% of total production value in 1986–88 to 51% in 2008 (Agricultural Statistics 2009). Around 2002, direct subsidies were about half of the income of the sheep production industry (Arnalds and Barkarson 2003).

During the last 100 years, Icelandic agriculture has had mixed success in maintaining the credibility and trust that it needs to maintain the public goodwill that underpins these strategic considerations. We can illustrate this by examples of farm forestry and grazing of private and common lands. In both sectors, public concerns for the environment have led to significant shifts in government support, which in turn have shifted the economic bases for farming activities.

Icelandic agriculture is characterised by high levels of governmental support, direct and indirect import barriers, an image of almost veterinary drug-free high quality produce, and a growing realisation of the need to maintain prosperous agriculture for food security. However, the immense societal change of the last decades in combination with debates over the role of governmental support in the immense land degradation experienced in Iceland has had a big impact on public goodwill towards agriculture. Maintaining a *social licence to farm* – that is, the credibility and trust that is required for prosperity – will be an increasingly complex task for Icelandic agriculture in the future.

The extensive land degradation that has, and is still, taking place in Iceland results primarily from unsustainable livestock grazing, traditionally by sheep, but also horses, in interaction with harsh climate and sensitive soils. Over the last century, agricultural policies have been a large determinant of land condition, firstly leading to the worst outcomes, but more recently as an emerging tool to attain sustainability of land use and maintain the social licence to farm.

Sheep farming in Iceland rests on deep cultural roots, shaped over many centuries by dependence on sheep to sustain a land-based domestic economy. In the 20th century, a political environment, or a social licence, existed for decades within which various agricultural organisations could influence policy on marketing and secure acceptable prices for their products.

In 1944, the newly founded Republic of Iceland found itself faced with economic problems. Investments under a state-controlled 'development programme' were therefore directed towards earning foreign currency and promoting domestic production. This included an incentives scheme with a guaranteed price for sheep. This resulted in expansion of sheep flocks, with little regard for the environment.

Sheep flocks peaked in 1978. This was followed by several years of exceptionally unfavourable growing conditions. The pressure, combined with declining vegetation, became more than the common grazing lands in the highlands and extensive pasture-lands in the lowlands areas could withstand. The power of soil conservation authorities was weak and there followed a period of extensive ecosystem damage.

Every country has its own sets of solutions for resolving land degradation problems and attaining goals of sustainable land use. Icelandic experience demonstrates the failure of top-down approaches and the importance of helping people conserve and heal the land. However, it also demonstrates the importance of community support for any particular farming activity, which can lead to generous arrangements to support that activity. Participation is now the main characteristic of soil conservation and land restoration in Iceland. Growing environmental awareness and success stories within the agricultural sector have had an increasing role in affecting public attitude towards agriculture and thus acceptance of the social licence to farm. This is particularly evident with the tasks of restoring land quality and support for farm forestry.

Restoring the woodland cover of Iceland has been a national priority for more than 100 years. The first task, supervised by the State Forest Service, was to prevent further destruction of the remaining woodlands. Then afforestation in treeless areas began, at first with the native birch, but later with emphasis on introduced species that grow well in Iceland.

Forestry has a long history of public participation in Iceland. The Icelandic Forestry Association (IFA) was founded in 1930, and is a national umbrella organisation for 57 local and regional forestry societies. These are non-governmental volunteer organisations of people interested in afforestation. The IFA has roughly 7000 members, or about 2.5% of the Icelandic population, making it by far the largest environmental NGO in Iceland. Strong social support is translated into favourable conditions for this farming activity.

Reflecting favourable public opinion towards forestry, state-supported afforestation programs have been established for all parts of Iceland. Five Regional Afforestation Projects established since 1991 were responsible for 80% of tree planting in Iceland in 2007. Close to 25% of Icelandic farms participate in such programs. Each farm afforestation grant covers 97% of establishment costs including fencing, roads, site preparation, planting and the first thinning. A farmer afforesting a large tract of land can earn what amounts to be as much as 2 to 3 months of wages per year (Eysteinsson 2009).

Despite the immense popularity of forestry in Iceland, several factors can affect the social licence underpinning continued support. Historically, forestry was at odds with traditional agriculture. Disputes between leaders of agriculture and forestry could be colourful. In many areas, owners of sheep are not required by law to keep their animals within enclosed farms, using adjacent properties and mountainous areas as commons. Forestry, which required fencing to protect trees from grazing, was seen to disrupt the free roaming of sheep, so was a threat to conventional livestock farming.

Forestry is now adapting to emerging community ecosystem and landscape concerns. As has also been experienced in Australia, large monoculture stands of introduced trees can greatly affect the beauty of unique (Icelandic) landscapes and be harmful for biodiversity and other ecosystem factors. Again, in response to shifting community concerns, afforestation goals have been changing from an emphasis of using introduced trees to an increased use of native birch and other species not grown for wood production. However, the main thrust of government supported forestry in Iceland may be regarded as single-issue, focused on timber-oriented farming instead of being seen as a tool to recover land quality and broad ecosystem values.

Social licence driving farmer-led conservation action

The farming community owns or has grazing rights to most of Iceland. Without their participation and commitment, sustainable use and restoration of land quality at national scales are unattainable. Since the all time record high in 1978, sheep numbers have halved, responding to economic and environmental realities. At the same time, the agricultural community has been taking an increasing role in improving land condition.

Flock reduction is largely a response to the loss of export subsidies and price guarantees from the government in 1991. Among social licence drivers of these changes was the public outrage over the great financial support the farmers were receiving, which drove costly overproduction and the resultant land degradation. The degradation problems were strikingly visual, and the media brought to the nation merciless coverage of grazing in areas of severe soil erosion. This created the perception that most of the land area of Iceland was affected by overgrazing and soil erosion. The government was increasingly seen as paying at both ends: indirectly for damaging the land by supporting too high livestock numbers, and directly for the costly reparation of land degradation stemming from the overgrazing.

Peer pressure started building up within the sheep industry, with a realisation that negative media coverage and increasing environmental awareness among the consumers might become very costly to the sector. Many sheep producers and leaders of agriculture began to understand the economic and regulatory implications of the potential loss of their social licence. They determined that their own initiative would be the best defence in maintaining the goodwill and trust required for maintenance of public and political support. This has enhanced adoption of the *Farmers Heal the Land* program that was initiated in 1990. Farmer support for this partnered-sustainability program has been a substantial contributor to rebuilding public trust in the farm sector.

The program was the result of contact in 1989 between members of the Icelandic SCS and the emerging Landcare movement in Australia. The Farmers Heal the Land program is locally led. Its purpose is to assist farmers to revegetate degraded land, halt erosion and reclaim land so that it again becomes available for sustainable use (Arnalds 2005a). It represents a change from the top-down approaches of the past, because it moves towards farmer participation in public good investment through caring for the land. It also demonstrates the way in which government financial support follows community attitudes to farming.

The government, through the SCS, refunds 85% of costs of fertilisers and provides seed where needed. The farmers contribute about 50% of the total project costs, mainly in kind through their machinery, skills and time required. This bottom-up approach encourages involvement and individual ownership of conservation and restoration projects. SCS staff act as technical advisers, facilitators and monitoring agents. This approach has been important in building trust between farmers and conservation authorities, which provides a foundation for resolving many other issues. The participating farmers have also been active in developing new methodologies, in cooperation with the soil conservationists, greatly advancing the knowledge base for soil conservation and land restoration.

The program has been a great success, but (as with the Landcare program that was its inspiration) there is much more to be done. Not all farmers have the willingness, or indeed the capacity, to embrace active conservation and restoration. About 25% of Icelandic farmers participate in restoring land quality through the Farmers Heal the Land program. According to Schmidt (1999), the three main reasons (75% of responses) for participation were: (1) to improve the visual appearance of rural areas; (2) environmental concerns; and (3) to deliver the land in better condition to the next generation. The main driver for participation may have been the visualisation of restoration of land quality as a tool to improve their social licence to farm. Though this program, the farmers have been able to demonstrate their crucial roles as trustworthy stewards of the land. At the 3rd Forum on the Future of Agriculture, in March 2010, the European Landowner's Organisation honoured a participant of the Farmers Heal the Land program with the 'Environment and Soil Management Award' for his outstanding work and innovation in restoring ecosystems on severely degraded land.

Agricultural support as an incentive for Landcare

The Farmers Heal the Land program has been an efficient tool for both improving land condition and raising conservation awareness among participating land users. This re-alignment of farming practice to fit evolving community awareness has also proven to be a sound strategic move. It has increased the visibility of the farming community as partners in achieving important environmental goals, such as restoring biodiversity and meeting national goals in mitigating climate change through carbon sequestration (Arnalds 2004). The land restoration role has helped in maintaining social justification for farming and aided in significant changes in support programs for agriculture.

The release of the first national survey of soil erosion in Iceland in 1997 (Arnalds *et al.* 2001) brought the debate on the severe land degradation problems from the 'I think' stage, to 'I know'. This assessment, involving formal verification of the often strikingly visual situations, had a profound impact on subsequent agricultural agreements, and formed the basis for new soil conservation programs.

In 2000, a new subsidy agreement was signed between sheep farmers and the government for the next 7-year period. All farmers with a production quota were entitled to the subsidies. A part of the contract was based on quality management, with gradual increase in government support to 22.5% during the contract period to farmers who met the quality criteria (Arnalds and Barkarson 2003). The criteria for quality management include good animal treatment, controlled use of chemicals and medicine, and participation in a national breeding program. Participation was on a voluntary basis, but almost all sheep farmers entitled to the governmental support choose to undergo these quality management criteria in order to receive the additional support.

A new contract between the Icelandic Government and the Farmers' Association of Iceland was signed in 2007 for improving the operational conditions of sheep production for the period 2008–2013. The objectives of the contract include strengthening sheep production as an industry and to ensure that sheep production is based on environmental conservation, land quality and sustainable land use. The proportion of support, based on participation in the voluntary quality management, was greatly increased. The governmental support includes direct payments to sheep farmers and indirect support through the Farmers' Association. The share of quality management is 52% of the direct payments, with around 27% of total support to sheep production.

The criteria for quality management are gradually being strengthened. Initially, making governmental support within the sheep farmer contract partially dependent on quality criteria met considerable opposition. Hence, the program began with fairly relaxed criteria with regard to land use, rather than risk losing the opportunity for subsequent achievements in the quest for sustainability. Key concepts include assistance with improvement of land quality and to give the farmers enough time to adjust to this new reality. On the sustainability issue, the participatory program Farmers Heal the Land had cultivated both the quality of the land and the mindset of many sheep farmers, some of whom have now been working on reclamation of their land continuously for 20 years in cooperation with the SCS. Farmers with land not passing quality criteria must submit a conservation and land improvement plan for approval in order to get the full subsidy. The Land Improvement Fund of the SCS has partially supported such schemes, especially on the common grazing lands.

The policy change to make governmental subsidies partially linked to voluntary participation in quality management has been a milestone in the quest for improved sustainability of land use in Iceland. In their analysis of soil erosion and land use policy in Iceland in relation to sheep grazing and government subsidies, Arnalds and Barkarson (2003) concluded that this linkage was brought about as a result of public pressure, partly triggered by the soil erosion assessment (Arnalds *et al.* 2001). This demonstrated in a systematic manner the poor condition of a high proportion of the area of the land used for sheep grazing. An important factor in the long-term success of this policy adjustment to sustainability is that it was brought about at the initiative of the farmers, determined to improve the image and the quality of their products, with the aim of securing the foundation for sheep farming in Iceland. This initiative, with quality of land management as one of the criteria defining their economic survival, is proving to be highly successful in strengthening the social licence for this important agricultural sector into the future.

Reflections on the journey

What does this history demonstrate about the social licence for farming? There is a pattern in the evolution of social and environmental responsibility of industry (including farming) that is well illustrated by the evolution of grazing incentives and controls in Iceland. It begins as farmers, in part responding to government stimulus, pursuing economic growth that puts at risk the underlying common resource. In the case of Iceland, this resource is soil and vegetation; in Australia it may be water. The community becomes aware of an emerging crisis of over-consumption of these natural resources, probably through the mass media. Farmers feel ill-treated, but some begin to reconsider. These thought-leaders provide the impetus for change within the industry.

Governments, also in response to perceptions of community sentiment, begin to realign regulation, education and incentives. Over time, the pattern of economic incentives and regulated controls begins to shift to better reflect a more constrained view of the freedom of users in resource exploitation. For some farmers, the transition may be difficult, but for others (already concerned about the condition of their land), making the transition is emotionally fulfilling, even if it sometimes economically difficult. The role of government is to support those who are keen to realign their practices to better fit their current and future resource needs and the changing community expectations, and (eventually) to force those who are not willing to adjust to the boundaries set by society.

This change process is, however, neither as simple nor unemotional as this description might suggest. It will need a clear vision of the future state of landscapes and ecosystem resources in Iceland, and how these are matched to both farmer and community needs. Maintaining a *social licence to farm* – that is, the credibility and trust that is required for prosperity – will be an increasingly complex task for Icelandic agriculture in the future. This never-ending journey involves conflict, stress, joy and hope in the attempts to achieve the desired milestones. It is important for the farming community to be seen as stewards of the land they have in trust for the next generations and to become active partners in designing and carrying out the policies, strategies and programs marking this route. It is a truly human journey, and a reflection of both the goods and the ills of democracy.

References and further reading

Agricultural Statistics (2009) The 'Farmers' Association of Iceland'. <http://bondi.is/lisalib/getfile.aspx?itemid=2211>.

Arnalds A (1987) Ecosystem disturbance in Iceland. *Arctic and Alpine Research* **19**(4), 508–513.

Arnalds A (2004) Carbon sequestration and the restoration of land health. *Climate Change* **65**(3), 333–346.

Arnalds A (2005a) Approaches to landcare – a century of soil conservation in Iceland. *Land Degradation & Development* **16**, 113–125.

Arnalds A (2005b) Barriers and incentives in soil conservation – experiences from Iceland. In *Strategies, Science and Law for the Conservation of the World Soil Resources*. pp. 251–259. Agricultural University of Iceland Publication No. 4, Borgarnes.

Arnalds A and Runolfsson S (2009) Iceland's century of conservation and restoration of soils and vegetation. In: *Soils, Society & Global Change*. (Eds H Bigas, GI Gudbrandsson, L Montanarella and A Arnalds) pp. 70–74. Soil Conservation Service of Iceland, Reykjavik.

Arnalds Ó, Thorarinsdóttir EF, Metusalemsson SM, Jónsson A, Grétarsson E and Árnason A (2001) *Soil Erosion in Iceland*. Soil Conservation Service and Agricultural Research Institute, Reykjavik.

Arnalds Ó and Barkarson BH (2003) Soil erosion and land use policy in Iceland in relation to sheep grazing and government subsidies. *Environmental Science & Policy* **6**, 105–113.

Douglas M (1996) A participatory approach to better land husbandry. In *Soil Conservation Extension from Concepts to Adoption*. (Eds DW Sanders and MG Cook) pp. 107–121. Soil and Water Conservation Society of Thailand, Bangkok.

Eysteinsson T (2009) Forestry in a treeless land. <http://www.skogur.is/english/forestry-in-a-treeless-land/>.

Hannam ID (2000) Soil conservation policies in Australia: successes, failures and requirements for ecologically sustainable policy. In *Soil and Water Conservation Policies and Programs: Successes and Failures*. (Eds EL Napier, SM Napier and J Tvrdon) pp. 493–514. CRC Press, Boca Raton, Florida.

Jonsson A (2009) *Why Iceland?* McGraw Hill, New York.

Olgeirsson F (2007) *Sáðmenn sandanna* [A Centennial of Soil Conservation in Iceland]. Soil Conservation Service of Iceland, Reykjavik.

Roberts R (1989) 'Land conservation in Australia: a 200 year stock-take'. Soil and Water Conservation Association of Australia. West Pennant Hills, NSW.

Sanders DW (2000) The implementation of soil conservation programmes. In *Rangeland Desertification*. (Eds O Arnalds and S Archer) pp. 143–151. Advances in Vegetation Sciences Series. Kluwer, Dordrecht, The Netherlands.

Schmidt G (1999) Bændur græða landið [Farmers heal the land]. BSc thesis. Agricultural College of Iceland [In Icelandic], Hvanneyri.
Thorsteinsson I (1986) The effect of grazing on stability and development of northern rangelands: a case study of Iceland. In *Grazing Research at Northern Latitudes.* (Ed. O Gudmundsson) pp. 37–43. Plenum Press, New York.

9

American agriculture's social licence to operate

John Becker and Amanda Kennedy

This chapter proposes to answer the question, 'What is the status of American agriculture's *social licence* to operate in the 21st century?' The chapter begins with reference to Thomas Jefferson, whose views on a wide range of social, political and economic aspects of the American democracy profoundly affect it. This discussion includes a historical summary of the US agricultural economy and the impact it has on the social licence of agriculture over time. Access to open land and opportunities to acquire ownership of it reflect the importance of private property to the US political and economic system. We reflect how land ownership in an agriculturally based economic system was an important means to acquire political influence locally and regionally. With the historical perspective providing our background, attention then turns to modern examples. These support the conclusion that the social licence of American agriculture to operate is supported by a variety of policies. Although these policies were established to support an agricultural economy of a particular type, the agricultural economy has changed. Intersecting notions of the family farm, corporate-scale farms and industrial-scale farming are important concepts to understanding how American policy makers view modern agriculture. The chapter concludes with an examination of trends that raise questions about continued support for American agriculture's social licence. Agricultural production is evolving in dramatically different directions, which raises the question: 'Is American agriculture's social licence to operate in jeopardy as a result of its response to economic pressure to complete on a global scale?' This question is of relevance to more than the US farming community. Similar historical characteristics and evolution from artisan to industrial farming exist in many other countries. Similar threats to the social licence upon which farmers have traditionally relied are emerging, and new strategies will be needed to justify support policies in the eyes of a sceptical public.

Thomas Jefferson's Agrarianism and its relationship to democracy and democratic development

Thomas Jefferson was a public official, philosopher and plantation owner who enjoys a special place in American history. As the author of the *Declaration of Independence* and the country's third President, the first elected from an opposition political party, Jefferson helped to shape the American democracy in many ways, including helping to shape the American view of agriculture.

In the mid 18th century, a group of French economists known as 'Physiocrats' promoted the view that the wealth of nations was based solely on the value of its land used in agriculture or in other development.[1] Cities were derided, while farmers and others in rural communities were lauded. Thomas Jefferson wrote widely on the place of farmers and farming in the new democracy without adopting the views of 'Physiocrats', who favoured large-scale farming and great estates (Griswold 1946). Jefferson cast his support in favour of frontier farmers and viewed agriculture not as a source of wealth, but as a source of virtue that supports self-government. From Jefferson's broader writings have been distilled principles that are often described as 'Jefferson's agrarianism' (Dixon and Hapke 2003). To Jefferson, agrarianism consisted of five core principles and beliefs: (1) a belief in the independence and virtue of the yeoman farmer; (2) the concept of private property as a natural right; (3) land ownership without restrictions on use or disposition; (4) the use of land as a safety valve to ensure justice in the city; and (5) a conviction that with hard work anyone could survive in farming. Independent landowning farmers were presented as pivotal to creation and maintenance of a democratic society.

Scholars observe that Jefferson's belief that farming had a moral and ethical primacy over industry firmly became part of America's ideological framework (Dixon and Hapke 2003). Towards the end of the 19th century, the family farm, then a predominant form of American agriculture, was at the centre of American society. Jefferson's emphasis on social recognition of farmers, their way of life and the importance of private property to America's political and economic system added emphasis.

The impact Jefferson's views had on American public policy can be seen in many examples, including: the *Morrill Land-Grant College Act of 1862*, which created the land grant college system for research and teaching in the agricultural and mechanical arts; the *Agricultural Adjustment Acts* of the 1930s and '40s; and the various comprehensive federal legislation from the 1940s to the present that were designed to influence agricultural production, protect farm income, and more recently, pursue environmental protection and conservation objectives (Dixon and Hapke 2003). Subsidies to agriculture were not considered as with welfare payments, but rather were viewed as investments to support continuation of this sector of the economy (Dixon and Hapke 2003). In addition to these federal measures, individual states passed laws to create farmland preservation programs that purchased the right to convert land from agricultural use to development, and thereby presented the loss of farmland, established preferential assessment of agricultural lands for local real estate taxes (reducing *ad valorem* values used to calculate annual real estate taxes), create preferential treatment (including full exemption) under land use planning laws, and grant special treatment under state imposed income or wealth transfer taxes.[2] In the following pages, attention will be paid to state laws designed to grant immunity to agricultural operations from nuisance suits brought by neighbours who challenge the right of a producer to operate on the basis that the production creates substantial interference with the neighbour's use, possession or enjoyment of the neighbour's own property.

Jefferson's agrarianism is not without its critics, including one who describes it as more fiction than fact and one that offered benefits to men, but only hard work, sacrifice and subservience to white and black women, whether slave or free.[3]

A historical view of agricultural activity in the US

People coming to the colonies were motivated by the desire to gain wealth, opportunities for jobs and availability of open land.[4] Initially ownership of land followed the European tradition of sovereign ownership and control over the known or discovered territory. As new discoveries occurred, sovereign holdings increased. Indigenous people were confronted. Where peaceful coexistence was not possible, conflict ensued with Indigenous people suffering the greater loss in these conflicts. Although land was available, after Indigenous claims were resolved ownership was limited to those who had been given royal favour to settle the New World.

Those who survived the journey to the New World were adventurers, rather than farmers (Albrecht and Murdock 1990). They soon learned that life in the New World revolved around agriculture, first at subsistence levels and later to gain income as a result of trading (Cochrane 1993). Through land ownership, individuals gained political as well as economic power. As communities formed to support agricultural enterprises, land ownership was the means to power and influence in the community and fledgling democracy.

In 1850, the US Census counted nearly 1.5 million farms in the United States Cochrane 1993). As westward expansion occurred in the later half of the century, land availability became a tool of government policy to settle the western territories.[5] The number of farms grew dramatically until it peaked in 1920 at more than 6 million farms (Albrecht and Murdock 1990). Currently there are slightly more than 2.2 million farms in the US, but less than 1% of the population describes its occupational status as farmer or farm manager.[6] An additional 950 000 people describe their occupation as farming, fishing or forestry.[7]

The percentage of US land in farms peaked in 1950 at more than 50% (Albrecht and Murdock 1990). Currently there is approximately 40% of US land in farms. In 2007, the average farm size was 418 acres, which is a slight decline in average farm size over the last 30 years.[8] In terms of the proportion of total US population engaged in farming, the number of people living on farms in 1880 was determined to be 43.8% of the total population.[9] From that point the percentage declined to its current level, which is less than 2% of total US population.[10]

Perhaps the greatest change in the face of US agriculture has taken place since the end of World War II. Technological breakthroughs in the mechanisation of production and widespread adoption of chemical fertilisers enabled people to leave labour-intensive agricultural production for manufacturing jobs in the growing cities across the country (Albrecht and Murdock 1990). Agricultural productivity also soared during this period. In 1900, an average farm worker produced enough food and fibre to supply seven other people. By 1990, that number of people who could be supported by the labour of one farm worker grew to 120 people (Albrecht and Murdock 1990).

From its inception, the family farm exemplified the organisational structure of a production enterprise. It was the family who were the source of labour, management and capital needed to manage the farm operation on which all relied. As productivity increased through technological advances, this labour was free to explore other opportunities. Although changes have occurred over time, little impact has been felt on the organisation structure. Sole proprietorships continue to be the most common organisation structure of farm units across the country.[11] Partnerships and limited liability structures emerged in recent years as structures for mid to large-scale operations. These organisation structures provided a tool for continued development of the farm business by adding new members to it.

Although the number of farms and farm workers declined, aggregate agricultural production grew significantly as a result of technology and consolidation of farm production into larger scale production units. In current terms, US agriculture can be described as having a bi-polar structure, characterised by a large number of small producers whose overall contribution to the farm economy is modest, and a small number of large-scale producers whose contribution to the overall economy is considerable (Albrecht and Murdock 1990). In the past 30 years, the most dramatic changes have occurred in the category of farms producing from $50 000 to $99 999 in average annual sales. This class experienced more than a 50% decline in farm numbers over this period. Conversely, the number of farms in the category of annual average farm sales from $100 000 to $499 999 experienced an increase of about 20% in farm numbers over that time. The number of farms with annual average farm sales of more than $500 000 grew more than five fold in the same period.[12]

In terms of its social licence to operate, American agriculture is a key component of the national economy. Since 1950, the decline in the number of farms and the number of people engaged in agricultural production is significant. How did the declining number and increase in farm size affect agriculture's social licence? Could agriculture rely on its traditional status or would its relative decline result in the loss of social acceptance and status? In the discussion that follows we will look at 'right to farm' laws, which provide special protection of farmers against litigation that could require them to limit their farming activities, or compensate neighbours for harms arising from farming. This economic policy measure developed in the same period as American agriculture evolved into larger scale activities. We will use it to shed light on the question of how the structural evolution affected the social licence of American agriculture.

Public policies that exemplify American agriculture's social licence to operate

As residential development moved out of the cities and into the country, it was inevitable that conflict between residential landowners and established farm operators would occur. The farmers' social licence to operate was being challenged. Cases such as *Pendoly v Ferreira*[13] provide a good example of the situations that set the stage for adoption of laws intended to protect farm owners and operators facing complaints from their neighbours who found the smells, odours and dust associated with farm production practices to be disturbing. Although it was Ferreira in this case who first established his agricultural facility, the Court reached a conclusion favouring the objectors, despite evidence showing that Ferreira operated his facility with reasonable care. The message to Ferreira was clear. Despite his best efforts to operate his facility in compliance with established standards using reasonable care, that effort was not enough to withstand the challenges of people who came to the area after Ferreira's use was established. There was nothing that the farmer could do, or have done, to avoid being forced to significantly restrict his agricultural enterprise and forced to suffer the consequential financial loss. From cases such as this developed recognition that a legislated response to strengthen agriculture's social licence was needed in the face of increasingly frequent complaints from neighbours, some of who were farmers themselves.[14]

Legislation that provides public support for agriculture's social licence to operate is referred to as 'right to farm' or 'protection of agricultural operations from nuisance suits' laws (Centner 2006).

These laws often recite that their legislative purpose is to conserve, protect and encourage the development and improvement of agricultural land and to reduce the loss of agricultural resources by limiting the circumstances under which agricultural operations may be the subject of nuisance suits brought by neighbours or public nuisance ordinances passed by local government.[15] This reflects the historical position that farming is considered to be a culturally important way of life and the risk of losing it would be an unacceptable American tragedy (Dixon and Hapke 2003). Each of the 50 states enacted one or more of these laws. Specific terms vary in the degree of protection these statutes offer.[16] Despite the common use of the term, 'right to farm', these statutes do not grant an absolute right to engage in farming practices. Rather they are grants of conditional protection to agricultural producers who operate their facilities under terms that grant them protection described in the law.

Most examples of these statutes offer two general types of protections. On the municipal government side, local governments agree to encourage the continuity, development and viability of agricultural operations within their jurisdiction. Every municipality that defines or prohibits a public nuisance is to exclude from the public nuisance definition any agricultural operation conducted in accordance with normal agricultural operations or those reasonable and prudent activities that farmers adopt from time to time[17] so long as the agricultural operation does not have a direct adverse effect on the public health and safety.[18]

These laws also offer protection from private suits by providing that no nuisance action will be brought against an agricultural operation that has lawfully been in operation for one year or more prior to the date of bringing such action, where the conditions or circumstances complained of as constituting the basis of the nuisance action have existed substantially unchanged since the established date of operation and the activities are normal agricultural operations.[19] To accommodate future growth and development in the agricultural activities, these laws provide that if the physical facilities of agricultural operations are substantially expanded or substantially altered, protection is afforded if the expanded or altered facility is in operation without complaint for some period of time after the expansion or alteration occurs, usually 1 or 2 years.

Most laws of this type establish situations in which the protection can be lost. For example, agricultural operations that have a direct adverse effect on the public health and safety, that violate federal, state or local regulatory requirements or that are operated in a negligent manner may lose nuisance suit protection.[20]Although these laws reaffirm the social licence of agriculture, they often set parameters within which the protected activity must operate to gain the desired protection. For example, an agricultural operation that operates in a negligent manner or that directly injures someone through its activities is unlikely to enjoy the legal protections afforded by this legislated social licence. These conditions become the standards under which an agricultural producer must operate to gain advantage from the social licence protection. Setting a specific period within which an activity must operate without complaint before protection is triggered is similar in principle to a statute of limitation period, which conditions a person's right to bring suit by a requirement for timeliness in pursuing the claim or be barred from doing so if tardy.

American agriculture's social licence to operate in the 21st century

Despite the emergence of nuisance litigation, the relationship between American farmers and their surrounding communities has traditionally been close. As farmers prospered, communities sprang up to offer services and support for the agricultural economy. The converse is also true. As the fortunes of agricultural producers waned, so did the viability of communities that only survive if agriculture is successful. Sociologist Walter Goldschmidt in the 1970s looked beyond these considerations of mutual economic dependency and struck a different chord when he posed the following hypothesis: the quality of life in communities surrounded by small farms is superior to the quality of life in communities surrounded by large farms and, therefore, an increase in the scale of agricultural production will have a negative effect on the quality of life in that community (Goldschmidt 1978). Although somewhat reminiscent of Jefferson's agrarianism, this hypothesis was posed at a point when the number of American farmers was in decline and the size of American farms was growing through consolidation and productivity gains. Scholars have questioned if this hypothesis could be generalised beyond the situations that Goldschmidt examined. Others, although pointing out its shortcomings, have concluded that the hypothesis can be supported in other applications (Peter 2002). The Goldschmidt hypothesis raises additional ammunition to fuel the debate between family-scale agriculture and large- or industrial-scale agriculture. Not only do these entities compete in the market place for the goods they produce, but they may also cause very different impacts on the communities that surround them. This hypothesis indicates that the benefit achieved by a community from surrounding agriculture may be inverse to the scale and economic efficiency of the farms. This in turn suggests that the justification for 'protecting' these farms from other claims against them reduces with the scale of the farming activity. In other words, the more industrial the farming activity, the less objective justification there is in sheltering them from the pressures that are imposed on industrial firms generally.

A debate has emerged questioning whether only one form of agriculture is worthy of having a special social licence to operate, or whether modern agriculture as it has evolved is deserving of social support. The answer to this question is important. If justification for social licence can be lost through growth and expansion, arguments made need to further sharpen the terms under which agricultural social support applies.

Despite public policy measures adopted to protect agricultural production, there have been instances where the loss of that support has become apparent. As agriculture has changed, questions are raised about the type of agricultural activity that is entitled to receive protection. An example is found in a Kentucky State Attorney General Opinion[21] that considered whether Kentucky's 'right to farm' statute prohibited county level municipal governments from regulating industrial-scale hog-raising in the state. Under the Kentucky statute, local government was prohibited from regulating agricultural operations, which operated according to 'generally accepted, reasonable, and prudent methods of operating a farm to obtain a monetary profit; that comply with applicable laws and administrative regulations, and is performed in a reasonable and prudent manner customary among farm operators.' Focusing on the final words of the definition, and citing the example of a manure lagoon failure at a large-scale livestock facility in North Carolina that discharged millions of gallons of untreated hog manure and drowned animals,[22] the Attorney General concluded that industrial-scale production facilities: '... hardly deserve to be called a farm at all. An industrial-scale hog operation is less a farm than a manufacturing facility' and that establishment of these facilities have drawn '... a high level of community opposition to these massive hog operations.'

Fearing that a large-scale pollution discharge could be repeated, the Attorney General concluded that industrial-scale hog operations were not reasonable, prudent and accepted farming methods that would be protected by the state 'right to farm' law. This view suggests that, as farming evolves into a more technology and capital-intensive activity (whose scale of potential harm-doing commensurately increases) in the eyes of the community, it becomes something else. It loses its special social character, and becomes 'merely' commercial.

Joining the Kentucky Attorney General are a series of cases that have focused on key terms of state 'right to farm' statutes on which a coverage decision could hinge,[23] or by invoking other legal principles that could overcome the sweeping grant of immunity from nuisance suit. An example of such an alternative approach is the tort law concept of anticipatory nuisance, which would allow a complainant to pursue a suit before the nuisance activity takes place, because the likelihood of harm is so great that action to enjoin it must be taken to prevent greater harm. Although few US states have embraced the concept of anticipatory nuisance,[24] it is a tool that can overcome whatever protection 'right to farm' laws can offer by preventing the establishment of the facility and the application of immunity from nuisance suits.

Conclusion

The focal point of this chapter has been to review policy measures protecting American agriculture's social licence to operate. As the above discussion shows, elements of social licence derive from a deep historical and cultural tradition and have had wide and diverse impacts on American public policy. By late 20th century, however, changes in American society, as well as in American agriculture, merged to present challenges to these traditional views. Policy options had to be fashioned to respond to the challenges. One of these was 'right to farm' laws. Now these measures have also been the subject of challenges on grounds that are fundamental to the American political and economic system.[25] Although this predicament is lamentable to many, the drive to increase scale and gain greater productivity, which seems to have undermined agriculture's 'special case' for protection was driven by economic considerations. What is concerning is a bifurcation of views about the role of farming and the justification for a special social licence protection. Arguably, this is reasonable in relation to the very strong agro-industrial enterprises, which would seem to be more than capable to managing and defending their own interests. However, the more traditional, smaller scale farms may well suffer substantially if 'right to farm' protections are taken away totally.

What this chapter highlights is the way in which social licence issues co-evolve with social philosophy, economics and industrial development. It also has suggested that as farming has segmented in industrial scale and technological sophistication, its relationships with community have also segmented. Social licence arrangements that evolved prior to this segmentation are beginning to adjust to a new reality. This suggests that strategies and policies around the social licence to farm will have to become far more nuanced if they are to suit these new conditions.

Endnotes

1 <http://homepage.newschool.edu/het//schools/physioc.htm>.
2 Some examples of these measures include preferential assessment laws in many states that lower the value of property for tax purposes from fair market value to a lower value, which represents use of the land in agriculture. In a number of states, local government authority to control land use of agricultural property does not extend to agricultural land. The final example of preferential treatment of income or wealth transfer taxes include a variety of tax provisions designed to lower the value of property subject to estate or inheritance taxes or to treat particular items of property under various income tax measures.
3 Dixon DP and Hapke HM (2003), p. 161, citing Hurst RD (1994) *American Agriculture: A Brief History*. Iowa State University Press, Ames.
4 Many American history texts would support this statement. For example, see: <http://www.sitesalive.com/hl/hstg/private/hlhstgreasons.html>.
5 *The Homestead Act of 1862*, May 20, 1862.
6 2007 US Census of Agriculture, Table 1.
7 Ibid.
8 2007 US Census of Agriculture, Table 1.
9 2000 US Census, Occupations 2000, Table 1.
10 Ibid.
11 Census of Agriculture, 2002, Table 58 for current figures. See earlier census data for transition issues over time.
12 2007 US Census of Agriculture, Table 1.
13 345 Mass. 309, 187 N.E. 2d 142 (1963). Alfred Ferreira began to raise pigs in Topsfield, Massachusetts in 1949. After a few years his business prospered and the size of the operation grew to a level where it was twice as large as it was 10 years before. At about the same time as the Ferreira's pig farm was starting, residential development began in the Topsfield area. Ten years later, the new residents were complaining about the stench from the pig farm. Ferreira operated a quality pig farm and it was considered to be in the top 5% to 10% of pig farms in the area as far as quality of operation is concerned. When the neighbours filed a nuisance complaint, Ferreira defended his business on several grounds, including the fact that his operation existed in the local community before the residential development occurred and his operation and he did not operate it in a negligent manner. Weighing these arguments, the Court found that [a] course of conduct that would have been without

fault in a rural area, has, with the change in the environs of the farm to a residential district become unreasonable. Because he could not be expected to operate the pig farm in a way that would correct the nuisance complaints, the decision should be to enjoin continuation of the pig farm at the present location. Despite the fact that Ferreira was likely to suffer significant losses in selling the property, the Court upheld the injunction and allowed Ferreira a period of about 15 months to close the facility.

14 See Herrin v. Opatut, 248 Ga. 140, 281SE.2d 575 (1980) and Laux v. Chopin Land Associates, Inc, 550 NE.2d 100 (1990).

15 Taken from the Pennsylvania statute, 3 PA Cons. Stat. Ann. section 951 (2009).

16 For example, under the Pennsylvania 'right to farm' law a key term is *Normal Agricultural Operations*, which is defined as the activities, practices, equipment and procedures that farmers adopt, use or engage in the production and preparation for market of poultry, livestock, and their products and in the production, harvesting and preparation for market or use of agricultural, agronomic, horticultural, silviculture and aquaculture crops and commodities and is: (1) not less than ten contiguous acres in area; or (2) less than ten contiguous acres but has an anticipated yearly gross income of at least $10 000. The term includes new activities, practices, equipment and procedures consistent with technological development within the agricultural industry. If an activity is considered to be a normal agricultural operation, it can expect to be protected from nuisance suits if other conditions in the law are met. Note the reference to new activities, practices and equipment, which allows for growth and development of agricultural practices while still qualifying for protection.

17 3 Pa.C.S.A. section 951 et seq. (2009).

18 See Ibid section 953(a) (2009). See Ibid section 954(a) (2009).

19 See Ibid section 953(a) (2009).

20 KY OAG opinion 97-31, August 21, 1997.

21 Ibid.

22 For example, see *Herrin v. Opatut* and *Laux v. Chopin land Associates, Inc.*, note 22.

23 For example, see *Superior Farm Management, L.L.C. v. Montgomery* 270 Ga. 614, 513 S.E.2d 215, 1999 Ga. LEXIS 251, 99 Fulton County DR 926 (1999), which represents use of a somewhat traditional strategy to deal with a contemporary issue. Opponents seek to confront an adversary in settings where the adversary's strengths are neutralised. Owners of adjacent land filed suit against Superior Farm Management, L.L.C. that proposed to own and operate a 1345 acre hog breeding facility in Taylor County, Georgia. The property owners filed suit seeking injunctive relief to stop construction of the facility before it was completed and began operations. Under Georgia statutory law, where the consequences of a nuisance about to be erected or commenced will be irreparable damage and such consequences are not merely possible but to a reasonable degree certain, an injunction may issue to restrain the nuisance before it is completed. Mere apprehension of injury or damage is insufficient, but where it is made to appear with reasonable certainty that irreparable harm and damage will occur from the operation of an otherwise lawful business amounting to a continuing nuisance, the Court in *Superior Farm Management* held that equity principles can enjoin the construction, maintenance or operation of lawful businesses before they are established.

24 Is a grant of immunity a taking that requires payment of just compensation? – *Bormann v. Board of Supervisors in and for Kossuth County* 584 N.W. 2nd. 309 (Iowa, 1998) cert denied 525 US 1172 (1999). The Iowa Supreme Court decision in *Bormann v. Board of Supervisors in and for Kossuth County* is noted because of important issues it raises about the interpretation of 'right to farm' laws from a constitutional perspective. As in the case of most litigation involving these laws, the specific terms of the statute are particularly important. The case

involved a facial challenge on constitutional grounds to section 352.11(1)(a) of the Iowa Code that provides for creation of agricultural security areas and was brought by a group of landowners whose property was adjacent to an agricultural security area that the Supervisors created in accordance with state law. The Iowa legislature chose to offer a special incentive to agricultural landowners who placed their land in agricultural security areas by extending to them a special type of 'right to farm' protection, which would be in addition to any other type of 'right to farm' protection otherwise applicable. Succinctly stated, the Court's rationale for finding a taking occurred is when the County Board approved the application to create the security area it triggered section 352.11(1)(a). The approval gave the applicants immunity from nuisance suits. This immunity resulted in the Board's taking of easements in the neighbours' properties for the benefit of the applicants to do acts on their property, which, were it not for the easement, would constitute a nuisance. This amounts to a taking of private property for public use without payment of just compensation in violation of the federal and state Constitutions that prohibits the taking of private property for a public use without payment of compensation. In many countries, the sovereign's obligation to compensate a property owner for expropriating private property is recognised. In Bormann, the Court noted the legislature exceeded its authority by authorising the use of property in such a way as to infringe on the rights of others without the payment of just compensation. Other challenges to 'right to farm' legislation have not been as successful as the Bormann decision and it has not been widely adopted.

References and further reading

Albrecht DE and Murdock SH (1990) *The Sociology of American Agriculture: An Ecological Perspective*. Iowa State University, Ames, p. 42.

Centner TJ (2006) Governments and unconstitutional takings: when do Right to Farm Laws go too far? *Boston College Environmental Affairs Law Review* **33**(1), 87–148.

Cochrane WW (1993) *The Development of American Agriculture*. The University of Minnesota Press, Minneapolis, p. 22.

Dixon DP and Hapke HM (2003) Cultivating discourse: the social construction of agricultural legislation. *Annals of American Geographers* **93**(1), 142–164.

Goldschmidt W (1978) *As You Sow: Three Studies in the Social Consequences of Agri-business*. Montclair NJ; Allanheld, Osum and Company, New Jersey.

Griswold AW (1946) The agrarian democracy of Thomas Jefferson. *The American Political Science Review* **40**(4), 667.

Peter DJ (2002) 'Revisiting the Goldschmidt Hypothesis: the effect of economic structure on socioeconomic conditions in the rural Midwest'. Technical Paper P-0702-1, July, 2002.

10

Soil conservation in Europe

Luca Montanarella and Jacqueline Williams

Soil conservation in Europe started to gain political attention during the late 1990s, when the German government initiated a Europe-wide debate on the need for a common approach. The initiative, known as the *European Soil Forum*,[1] triggered a wide-ranging discussion at the European Union and national level in various European countries. During that debate, it became rapidly obvious that soil conservation was a complex issue, particularly in Europe, where long historical development has had a deep impact on European soil resources. European soils are the result of centuries of human interaction. They have been mostly shaped by the development of the various agricultural practices that have been adopted in Europe, starting with the first agricultural revolution of the Neolithic era. In Europe, there is very little soil left that may be considered as 'natural'. First agricultural, and later industrial, activities have had a large impact on the soils of the various landscapes of Europe.

This chapter discusses how Europe has translated *social licence* through institutions and laws to achieve soil conservation outcomes. We focus on the success of the EC Common Agricultural Policy (CAP) and recent case studies in the European Union (EU) on sustainable agriculture and soil conservation. This discussion provides insights into future directions in Europe for the social licence to farm.

The EU Common Agricultural Policy (CAP)

The CAP evolved from the 1950s within Western Europe. It was founded on the need to encourage better agricultural productivity to ensure a staple supply of affordable food and maintain a viable agricultural sector as a result of post-war policies. The CAP formally came into force in 1962. From the 1960s to the 1980s, the CAP achieved its main objective of securing food supplies through a range of subsidies, incentives and training. By the late 1980s, the EU was well on the way to achieving self-sufficiency. The success of the CAP, however, created perverse outcomes such as large surpluses of agricultural commodities, which resulted in dumping or storage within the EU and the need for export markets. These unanticipated outcomes resulted in high transaction costs, distortion of markets and concerns from consumers and taxpayers.

High-level media attention to the dumping of food within Europe created enormous political pressure to reform the CAP. In 1992, around the time of the Rio Earth Summit, concerns were expressed globally about the environmental sustainability of agricultural production and the food dumping that resulted from the EU subsidies. As a consequence, a 1992

CAP reform was implemented through a number of actions including: the lessening of agricultural prices to facilitate more competition in the internal and global market; compensating farmers for loss of income; creating new market mechanisms; and implementing measures to protect the environment.

In particular, this reform introduced the agro-environment measures, which encouraged farmers to provide environmental services through good agricultural practices. The first reform of the CAP was successful in reducing surpluses and controlling expenditure without compromising farm income. The EU, however, required a new model for the future of agriculture to ensure good practice and environmental protection, with an added focus of rural development. In 2000, the second CAP reform occurred. This introduced a new policy for rural development as the 'second pillar'. This reform included: increasing competitiveness; ensuring fair standards of living and employment opportunities for farmers; greater environmental and structural considerations; improved food quality and safety, and harmonisation of agricultural legislation and decentralisation of its application (European Commission 2007; European Commission 2010c).

The CAP reform of 2003 resulted in compulsory cross-compliance for all farmers receiving direct payments from the EU, which was extended to include compliance with environmental rules, including new requirements on public, animal and plant health, animal welfare and the maintenance of all agricultural land in good agricultural and environmental condition (known as the Good Agricultural and Environmental Conditions or GAEC). This is applied in each EU member state at national or regional level under a common agreed framework. The significance of the GAEC is that it applies to all direct income payments to farmers, including most environmental payments applied under Rural Development. Cross-compliance provides a baseline for the voluntary agri-environmental measures. Farmers are granted payments when they undertake environmental commitments that go beyond mandatory requirements. The GAEC is underpinned by a range of standards relating to protection against soil erosion, maintenance of soil organic matter and structure, avoidance of the deterioration of habitats, and water management (European Commission 2010a).

The CAP implementation of soil conservation

Recognition of the dramatic soil degradation processes occurring in Europe was formalised by the European Commission in 2006 through the submission of a formal communication (European Commission, COM(2006)231) to the European Parliament and the European Council outlining the *EU Thematic Strategy for Soil Protection*. The strategy is built upon four pillars being:

- a framework legislation (Soil Framework Directive) containing the binding elements for the strategy
- the integration of soil protection measures within existing and future legislation covering other areas related to soil
- research and development on soil protection aspects still lacking sufficient scientific knowledge to be immediately included in protection measures
- awareness raising initiatives at all levels to increase the general public awareness on the importance of soil protection for sustainable development.

The soil thematic strategy recognises the key role farmers' play in protecting soils in rural Europe. This role was enshrined in the 'mid-term' review process of the Common Agricultural Policy (CAP) of 2003. The CAP comprises two principal forms of budgetary expenditure: market support, known as Pillar One, and a range of payments for rural development measures, known as Pillar Two.

Cross-compliance, a horizontal tool for both pillars, which has been compulsory since the implementation of the CAP reform 2003 (Council Regulation (EC) No 1782/2003), plays an important role in soil protection, conservation and/or improvement. Under cross-compliance rules, the receipt of the single farm payment and payments for eight rural development measures under axis 2 is conditional on a farmer's compliance with a set of standards.

Enforcement of implementation and control of EU environmental directives were promoted through compliance with the Statutory Management Requirements (SMR). The GAEC were introduced to prevent land abandonment that could result from the decoupling of financial supports from production. GAEC specifically includes protection against soil erosion, maintenance or improvement of soil organic matter, and maintenance of a good soil structure. The fact that GAEC is defined at a national level enables member states to tackle soil degradation processes flexibly according to their own national priorities and local needs. Some member states used GAEC to compensate for gaps in their existing national legislation on soil protection, while other member states already had a legislative basis in place and merely adopted it for cross-compliance. This has resulted in national designs of GAEC that are highly variable in scope and detail.

Within the second pillar of the CAP, a wide range of measures can be supported under Council Regulation (EC) No 1698/2005. Member states and regions are obliged to spread their rural development funding across three thematic axes being: (1) competitiveness; (2) environment and land management; and (3) economic diversity and quality of life, with minimum spending thresholds applied per axis (i.e. 10% for axes 1 and 3, and 25% for axis 2). This brief history illustrates the extent to which policies to manage and support agriculture have evolved, through political processes, to reflect the perceived needs of the community. With changes in policy have come significant adjustments to farming practices.

An innovative EU program 'Leader', which was first introduced in 1991, is a key delivery mechanism of the CAP. The main concept of the 'Leader' approach is that strategies are more effective and efficient when designed and implemented at a local level by local actors, alongside clear and transparent procedures supported by public administrations coupled with technical assistance for the transfer of good practice. The 'Leader' approach is underpinned by seven key features (European Communities 2006) being:

- area-based local development strategies
- a bottom-up approach;
- public–private partnerships (through Local Action Groups)
- facilitating innovation
- integrated and multi-sectoral actions
- networking
- cooperation.

The European approach has been highly participatory, with substantial negotiation between farmers and government officials and a commitment to a partnership approach to regulation and support.

The 'Leader' program acts as a horizontal thematic axis (minimum spending of 5%; 2.5% in the new member states) complementing the three thematic axes. Axis 2 measures are of particular interest for those concerned with soil protection, because both environmental improvement and preservation of the countryside and landscape require management of soil degradation processes. In relation to environment and land management (Axis 2), member states are encouraged to focus on key actions. Some explicitly refer to soil, such as the delivery of environmental services, in particular water and soil resources, or the role of soils in adapting to climate change.

The full involvement of the relevant stakeholders, particularly farmers, is a pre-requisite for any successful soil conservation strategy. Therefore the European Commission, prior to the adoption of the EU Thematic Strategy, organised extensive stakeholder consultation for soil protection (European Commission 2004). The stakeholder consultation confirmed the important role of agriculture in soil conservation. Good agricultural practices are the key for reducing soil erosion, decline of soil organic carbon, soil compaction, salinisation and loss of soil biodiversity. Nevertheless, the responses to questionnaires during public consultation confirmed that the main concern of European citizens is still soil contamination by industrial activities (70% of respondents) and intensive agricultural practices (60% of respondents).

Sustainable agriculture and soil conservation

The important role of the CAP has been recently reinforced by a specific study launched by the European Parliament dealing with 'Sustainable Agriculture and Soil Conservation (the SoCo[2] project)'. In addition to improving knowledge on soil conservation in agriculture and the related policy framework, the project included dissemination activities targeting relevant stakeholders and policy makers across the EU. The project included 10 case studies within the EU, intended to acquire detailed information on the implementation of agricultural policies and measures, their effectiveness and consequences for soil conservation. This was compared with the monitoring and assessment undertaken on a continental scale (European Commission 2009a). Looking firstly at the SoCo project findings in a regional context, we will present the findings of the Italian case study followed by the overall project findings.

The Marche Region of Italy was selected for the Italian case study because of its geography. This region is geographically diverse, ranging from coastal areas to the Apennine mountain range: diversity common to many regions in Italy. Soil degradation, such as soil erosion, is widespread in Italy and the entire Mediterranean area, with common factors enabling the extrapolation of the results of the case study to a wider area.

Findings from the 'SoCo' research in the Marche Region concluded geographical aspects should be taken into account when defining policy and measures, with soil being an integral part of this geographical character. Extensive knowledge of soils, and of the delicate balance that maintains the soil's multiple functions, is essential for soil conservation. Conservation agriculture is best interpreted in relation to the local geography and the characteristics and quality of the soil, rather than being understood as a series of transactions (such as management practices).

Sustainable agriculture is best described as the achievement of a balance between the socio-economic and environmental factors. Environmental objectives are the cornerstone of the CAP, with the first pillar of cross-compliance introducing a strong innovative element in environmental protection. The second pillar of the CAP fully respects the geographical aspect. It allows for local policies and regulations based on the characteristics of the local geography.

Soil conservation achieves multiple EU policy objectives such as *The Nitrate Directive, The Sewage Sludge Directive* and *The Water Framework Directive*. The effectiveness of current policies and measures should be assessed through a monitoring network. The farmer questionnaire results discussed below clearly show that, although farmers do perceive soil degradation as being present, their perception of the intensity of the problem ranges from low to medium. The main risks of soil degradation identified by farmers include erosion, decline in organic matter content and reduced water retention capacity (European Commission 2009b).

Overall, the findings of the SoCo project across the 10 case studies provide useful lessons for policy implementation to achieve soil conservation. Cross-compliance was found to contribute to establishing a common reference level for sustainable soil management, with rural development measures an important instrument for assisting farmers' transition to higher levels of soil quality. This research identified:

- that the scale of intervention needs to reflect the scale of degradation
- the importance of developing context specific solutions
- the need to target farms with substantial problems
- that information and advice is essential to raising awareness and supporting change
- the importance of stakeholder involvement.

There is a clear role for a mix of policy measures to be used. It is also important to consider the potential for unintended side effects of environmental policies. To ensure ongoing success, relevant policy measures need to be coordinated and specifically targeted to soil protection. From the government perspective, forcing compliance is costly. There is a need for cost savings through devolvement and sharing of responsibility. Industry requires 'green credentials'. This suggests the potential for mutual gain to both farmers and the government from a co-regulatory approach, which also allows farmers to obtain market recognition for actions, which go beyond mere compliance. A combined policy and industry intervention using co-regulatory approach has the potential to achieve soil conservation objectives that respect the multi-functionality of soil and match soil quality with society's demands (Louwagie 2010).

The future of the CAP and GAEC

The next CAP reform is due by 2013. A formal public consultation is to be undertaken in late 2010. Preliminary to this process, in April 2010, the Commissioner invited all interested EU citizens and organisations to join in a debate on the future of the CAP. This was undertaken through an online dialogue through a website (open for 2 months) where respondents could post their views around four key questions. Responses were invited from (1) the general public, (2) stakeholders and (3) think tanks, research institutes and others. The four key questions were:

- Why do we need a European Common Agricultural Policy?
- What do citizens expect from agriculture?
- Why reform the CAP?
- What tools do we need for the CAP of tomorrow?

The future CAP

Five thousand seven hundred submissions were received, which were summarised by an independent group of experts and writers. Twelve core themes emerged from this preliminary review process (European Commission 2010b) identifying that the EU should:

- take a strategic approach to CAP reform
- ensure that the CAP guarantees food security for the EU, using a number of tools to achieve this
- continue to push competition within European agriculture in a market context, underpinned by innovation and dissemination of research
- transform market intervention into a modern risk and crisis management tool
- recognise that the market cannot, or will not, pay for the provision of public goods and benefits: this is where public action has to offset market failure
- ensure the correct payment to farmers for the delivery of public goods and services will be a key element in a reformed CAP
- protect the environment and biodiversity; conserve the countryside, sustain the rural economy and preserve/create rural jobs; and mitigate climate change

- rethink the structure of the two support pillars and clarify the relationship between them
- make adequate resources available for successful rural development
- implement a CAP that is fairer to small farmers and less favoured regions and to new member states
- introduce transparency along the food chain, with a greater say for producers
- create fair competition conditions between domestic and imported products
- avoid damaging the economies or food production capacities of developing countries and help in the fight against world hunger.

These responses will inform the formal consultation procedures leading up to the 2013 reform of the CAP.

The GAEC approach, which underpins the CAP, has also recently been under review. A GAEC workshop was held in October 2010 in Rome. One hundred and twenty-four experts attended this workshop, with the main topics being: GAEC minimum requirements in member states; best practices for minimum requirements; controlling GAEC with remote sensing; and environmental effects of some GAEC minimum requirements. This workshop found that there was a need for increased involvement of member states and other reforms, which include: amendments and redefinition of requirements; models and scientific approaches; increased participation in GAEC events, and new legislation. Additional needs that were highlighted through the workshop included (European Commission 2010b):

- clarifying the meaning of GAEC, its purposes and GAEC baseline, Eligibility/GAEC/Land Parcel Information System, improved practices (for example scientific evidence of effectiveness)
- improving communication of the GAEC concept to EU citizens and the GAEC concept to farmers
- increasing advice
- sharing information and data from research.

Conclusion

Changes in the pattern and extend of economic support for farming reflect changes to EU community priorities, and attitudes towards farming. Whereas previously the focus was upon food security and the social well-being of rural areas, increasingly support and regulation are being used to advance environmental values.

A full range of new European legislation is being developed to deal with industrial pollution within a range of environmental contexts, including protection of soils. Provisions within the CAP concerning intensive agricultural practices encourage extensivation[3], set-aside and organic farming through the use of a number of incentives. Nevertheless market-driven demand for agricultural commodities is often powerful, and further shifts of large areas of Europe into very intensive agricultural systems. Recent identification of areas of high natural value farmland should form the basis for measures to protect fragile agricultural landscapes, which contain a large pool of biodiversity to be protected for future generations.

There is a long way to go to fully empower European farmers as the main actors for achieving soil conservation goals. The recognition of farmers as providing not only agricultural commodities but also environmental services to all citizens needs still to be fully acknowledged.

The large CAP budget, which benefits a relatively small fraction of the population, is constantly under scrutiny. With an overall amount of approximately 50 billion €, the CAP absorbs around 40% of the EU budget (as compared with 60% in 1989). By 2013, this share is projected to fall to about 35%. Whereas, 10 years ago, 0.5% of the EU's GDP was spent on supporting EU farmers and rural areas, that figure now stands at 0.40% and is expected to drop to 0.33% in 2013; that is, less than 1% of total EU public expenditure.

As in other parts of the world, in Europe there is increased attention to ensuring that society gets tangible and measurable environmental improvement for its farmers. More effective accounting of the ecosystem services provided by farmers would allow for positive recognition of the contribution of the farming community by European citizens. A possible way forward could be the development of a close partnership between the farming community and the surrounding urban communities. Recent attempts at this approach appear highly promising, with organic farming; 'slow food', 'zero kilometre food' and similar initiatives bringing citizens closer to food producers, strongly contributing to the establishment of such partnerships. Such innovative partnerships deliver immediate gains in food quality, price to the consumer, environmental quality and social acceptance of the farmers and their activities.

'Healthy food from healthy soils' is a key message to achieve the reconciliation of the urban population with a farming community that is still perceived in many areas as the main cause of contamination and environmental degradation. Recognising farmers as the main guardians of the soils of Europe may pave the way towards a new era of participatory soil conservation strategies.

The recent 3rd Forum on the Future of Agriculture,[4] held in Brussels in March 2010, found an emerging consensus among policy makers and stakeholders that such a new role for European farmers has to be further developed and fully recognised within the future CAP. The strong call by the EU Commissioner for the Environment, Mr Janez Potochnik, for full integration of European agricultural and environmental policies into a single 'Common Agricultural and Environmental' policy is certainly giving a signal of the political direction the EU may be heading in the near future. Such integration will necessarily pass through the full recognition of sustainable land management as the key towards participatory soil conservation practices in Europe, involving the rural communities in close partnership with the urban communities. Several environmental policies will have to be streamlined with the agricultural policy in order to achieve a more effective impact on the quality of the environment. Not only soil, but water, air, nature and waste are policy areas that would benefit enormously from such a coordinated approach to land management and rural development. At the very end, placing the farmer's key role as the guardian of our landscapes and of the quality of what we eat, drink and breath will translate the social licence to farm into practice.

Endnotes

1 <http://ecologic.eu/download/projekte/900-949/943-944/944_european_soil_protection.pdf>.
2 <http://soco.jrc.ec.europa.eu/>.
3 'Extensivation' is a European term, the opposite of intensification.The term 'extensification' is also used.
4 <http://www.forumforagriculture.com/>.

References and further reading

European Commission (2004) *Environment – Soil European Commission*. Brussels, <http://ec.europa.eu/environment/soil/making_en.htm>.

European Communities (2006) 'The leader approach – a basic guide'. Office for Official Publications of the European Communities, Luxembourg.

European Commission (2007) 'CAP reform: implementing cross-compliance'. Office for Official Publications of the European Communities, Luxembourg.

European Commission (2009a) 'Case study – Italy sustainable agriculture and soil conservation (SoCo) project'. Office for Official Publications of the European Communities, Luxembourg.

European Commission (2009b) 'Final report on the project sustainable agriculture and soil conservation SoCo'. Office for Official Publications of the European Communities, Luxembourg.

European Commission (2009c) 'Requirement to keep land in good agricultural and environmental condition'. Office for Official Publications of the European Communities, Luxembourg.

European Commission (2010a) 'Main outcomes of the Rome 2010 GAEC workshop'. Joint Research Centre, Ispra.

European Commission (2010b) 'The Common Agricultural Policy after 2013 – Public debate. Executive summary of contributions'. European Commission, Brussels, <www.ec.europa.eu/cap-debate>.

European Commission (2010c) 'A history of successful change'. European Commission, Brussels, <http://ec.europa.eu/agriculture/capexplained/change/index_en.htm>.

Louwagie G (2010) Opportunities and limitations of soil policy for sustainable food production: the EU as an example of cases. In: *Conference on Sustainable Agriculture: The Art of Farming.* 11–12 May 2010, Brussels.

LEGAL AND INSTITUTIONAL ASPECTS

11

Social licence and international law: the case of the European Union

Jürgen Bröhmer

The A to Z of Corporate Social Responsibility (Visser *et al.* 2007) defines the concept of *social licence* in the context as:

> '... *the acceptance, express or implied, of a corporation's impact on people, society and the environment by their stakeholders or the public at large.*'

That, of course, is to distinguish the term social licence from the narrower legal meaning of licence, which refers to an authority to act or operate, or to use a good or right granted by the government or a rights holder. The definition proposed by Visser *et al.* (2007) is not very precise because the social licence concept cannot be restricted to the behaviour of a corporation because the form of ownership (privately held, incorporated or publicly held) has nothing to do with the social licence concept. Social licence applies equally to non-corporate users of resources as it does to companies. The important point in the present context appears to be the distinction of the social licence to operate from the sphere of legally defined rights and responsibilities.

The social licence concept (Visser *et al.* 2007) appears to be rooted in concepts of corporate governance and corporate social or environmental responsibility (CSR). CSR has become a blanket term for a range of issues to do with how visible institutions whose behaviour has an effect on the world around them should behave. Things discussed under this heading range across corporate governance and transparency to executive compensation, from industrial relations to forms of investment, and from consumer protection to environmental issues, to name just some of the areas relevant in this context.[1] A marketing or public relations element is evident, because many CSR issues have to do with improving the interaction between the institution in question and its social environment. That process of interaction does not appear to be entirely driven by exclusively altruistic motives. Similarities to, or overlap with, concepts such as branding or cause-related marketing are difficult to deny when engaging with the concept of social licence.

If one were to confine an analysis of the social licence concept strictly as a non-legal matter, it would make little sense to continue with this book. Although social licence is a socio-political matter, it is one that has substantial legal implications, and often it is given effect through the operation of laws. This chapter will attempt to argue that the two spheres cannot, and should not, be regarded as mutually exclusive zones and that the connection between the legal and political can be very close depending on the social, political, legal and institutional environment.

This chapter aims to use the European Union legal system to demonstrate the way in which legal and social interventions are linked, highlighting the ways in which politics marries these two elements. This is not to suggest that this link is in any way unique to Europe, because most chapters in this book suggest that the concept of social licence does, if nothing else, speak of this link.

This chapter shows a complex interaction between European laws that defend the social licence of the individual or the enterprise, and national laws intended to restrict this licence to meet particular needs of the state. It highlights an additional international law dimension to the social licence issue, which is generally discussed as being a dialogue between the nation state and its citizens. The role of supra-national organisations in defending the social licence of the enterprise and the citizen is an interesting modern development. The European instance highlights the potential for this supra-national dimension to come into being, and how this in turn can act as a brake on the ability of the nation state to restrict the commercial freedoms of its citizens. With the growth in international trade laws and other forms of convention that have supra-national effect, this dynamic is likely to be increasingly evident.

Social licence and legal frameworks

Tom Price commenced his observations on CSR with an example of a meeting of Chief Executive Officers' (CEOs) of major US firms in Washington to demand from the federal government regulation to restrict the emission of greenhouse gases.[2] The CEOs were asking for regulation of emissions that would be in the form of legislation, not on the basis of some other approach such as taxation penalties, or disentitlement to access to government programs. The preferred mechanism for imposing demands for change in corporate behaviour in a modern society is likely to be through the law. In a democratic society, it is hard to imagine that any form of social behaviour, corporate or otherwise, positive or negative, could not at least potentially attract the attention of lawmakers in the various parliaments, or the involvement of those who apply or develop the law in other capacities (such as citizen legal activists). That will at least be true if that behaviour is of public significance.

The concept of social licence appears to describe a process of communication between an institution and its social environment. The communication that takes place is by definition public and not secret. It is by its nature not limited either in its scope nor by the range of actions that are contemplated hence it can have legal implications as well as political significance. The social and legal spheres are in constant interaction mediated by the many communication processes that take place in an open, responsive and democratic society. Hence, when talking about the concept of social licence, the fact that one might primarily look for non-legal communication processes must not lead to a premature conclusion that this is not also a debate about the actual or potential role of the law.

The legal-institutional framework of the European Union from a social licence perspective

The European Union (EU) is perhaps better placed than most traditional state systems in providing opportunities to observe the relationship between legal regulation and the social licence concept. The EU is a state-like international organisation with far-reaching legislative powers. Its law is supreme to that of its 27 member states. Any legal norm of the EU overrides any norm of the domestic law of a member state, even constitutional norms.[3]

EU legislation is created through complex legislative processes involving the European Commission as sole initiator of legislation and the Council of Ministers as the representative body of the member states and the directly elected European Parliament as legislative bodies.[4]

The law thus enacted is not implemented by the EU's own administrative bureaucracy but by the bureaucracies of the member states and (not to be underestimated) by the sheer force of the rule of law. This is particularly by EU citizens taking up their cause in national courts on the basis of EU law empowering them and by the European Commission invoking the European Court of Justice in treaty infringement proceedings.

The fact that the EU does not have a bureaucracy to implement and enforce its own law is significant. The interests of the EU as expressed in its law are not always identical with the political interests of the day in the member states. Such disjunction does not only occur when a member state unsuccessfully opposes a certain legislative act. Disjunction is also possible when a member state supports a legislative measure at the EU level but finds it advantageous to behave differently at home. One reason for this can be sheer political opportunism. Another reason is that some member states are federations where the implementation and enforcement of EU law might be outside the scope of the jurisdiction of the federal government and lie with the constituent entities (for example the *Länder* in Germany and Austria or the autonomous regions in Spain) where a different political constellation might be in power.

In contrast to this political reality of the EU, the traditional nation state operates on the basis of largely congruent interests of government, the administration and the bureaucracy whose task it is to implement and enforce laws and regulations passed by the government of the day. As governments change, generally the focus of the congruent goals shifts with the change in political authority.

A second significant difference between the EU and a traditional nation state lies in the fact that the EU legal order is not only prescriptive order (it legislates and regulates matters) it is also a limiting, restrictive order. It places considerable limitations on what its member states can do 'at home'. It goes without saying that this will at times create political conflict. The common market is an illustrative example of this restrictive effect of the EU.

The result of these characteristics of EU law is two fold. Social licence considerations have an impact in the post-legislative political struggle over the enforcement of EU laws. Thus a state government in making decisions about the vigour with which it implements EU law will take into account considerations of the perceived social responsibility of the potentially affected corporations or citizens, given community norms within that state. Over and above this, social licence considerations are relevant for the EU itself in determining what rules it creates and how it goes about seeking their application within national states.

The common market of the European Union

The EU is described by the term *supra-national organisation*. The term was created to describe the phenomenon of a highly integrated international organisation that cannot be described as a state in the traditional sense, but is much more than a regular international organisation.[5]

The EU covers a wide array of policy fields, ranging from environmental policy to foreign trade, from coordination of social policy to justice, home affairs, foreign and security policy. The broadening of the scope of the EU was the reason for amending the name of one of the previous constituting treaties from *European Economic Community* to *European Community*.[6] However, the EU's common market has remained one of the cornerstones of European integration. The common market embeds four fundamental market freedoms:

- the free movement of goods (Articles 34–37 Treaty on the Functioning of the European Union (TFEU)[7])
- the free movement of services (Articles 56–62 TFEU)
- the free movement of people, consisting of

  - the free movement of workers (Article 45–48 TFEU)
  - the freedom of establishment (Article 49–55 TFEU)
- The free movement of capital and payments (Articles 63–66 TFEU).

The principal structure of the EU's common market regime

What these market freedoms do in effect is to provide a legal guarantee of the social licence of citizens to exercise these freedoms. They place this supra-national guarantee above national law, and provide a legal framework to enforce these freedoms against individual states. The legal approach governing the application of the fundamental market freedoms is similar for all market freedoms. For the purposes of this chapter, it will suffice to illustrate this using the example of the free movement of goods.

The starting point is the definition of the scope of the 'right' for free movement of goods. Because it is targeted at preventing state action that inhibits the movement of goods, rather than at controlling individual behaviour, the 'right' extends to protect the *social licence* of anyone moving or wanting to move goods within the territory of the EU, whether a EU citizen or not.[8] Article 34 defines the scope of the free movement of goods in a sweeping prohibition:

> *'Quantitative restrictions on imports and all measures having equivalent effect shall be prohibited between Member States.'*

The significant interpretative question is the meaning of 'measures of equivalent effect'. The European Court of Justice (ECJ) broadly interpreted this term as encompassing:

> *'... all trading rules enacted by Member States which are capable of hindering, directly or indirectly, actually or potentially, intra-Community trade are to be considered as measures having an effect equivalent to quantitative restrictions.'*[9]

The effect of this broad interpretation of the concept of 'measures of equivalent effect' means that almost everything member states do, or rules they have on their books with regard to market access, will fall within the scope of the prohibition and hence becomes subject to possible legal action based on the right of free movement of goods.

The ECJ in a later judgement limited the scope of the earlier judgement of *Dassonville* (EJC 1974) by introducing the concept of so-called 'selling arrangements'. In contrast to product rules, they are regarded as having no effect on the free flow of goods and hence are not limiting the scope of Article 34. Such 'selling arrangements' are, for example, laws about the opening and closing times of stores in the member states.[10]

Possible justifications for infringements

The social licence that is protected through these rights is not an absolute licence to move goods at any time or place, or without any constraint. In common with all other forms of economic interest protected by law, including private property, rights are subject to boundaries to protect other citizens' interests, and subject to legal constraints intended to protect the broader public interest. No 'right' is absolute and neither is the right for free movement of goods.

The TFEU provides for express exceptions in Article 36:

> *'The provisions of Articles 34 and 35 shall not preclude prohibitions or restrictions on imports, exports or goods in transit justified on grounds of public morality, public policy or public security; the protection of health and life of humans, animals or plants; the protection of national treasures possessing artistic, historic or archaeological value; or the protection of*

> *industrial and commercial property. Such prohibitions or restrictions shall not, however, constitute a means of arbitrary discrimination or a disguised restriction on trade between Member States.'*

What in effect this is saying is that there are forms of restriction on this particular social licence that can be seen as legitimate in the eyes of the international community (through the decisions of the ICJ), even if these cannot be definitively pre-specified. These exceptions provide an interesting insight into issues of legitimacy of the interests of the state and the broader community in overriding the freedoms that commercial interests believe they ought enjoy.

The ECJ has always interpreted exceptions restrictively, adopting a philosophy of protection of the social licence of industry and the citizen against controls by the state. Hence, it is always difficult for a member state to argue the Article 36 justifications for a national measure that can be brought under the *Dassonville* formula. In addition, the grounds mentioned in Article 36 are not sufficient to cover the complex area of justified exceptions. Important topics are missing from the text, for example: state restrictions that may be imposed in the interests of environmental protection; and consumer protection, trade practices and many other potential public interests that might require recognition. As will be shown below, these public interests can be closely tied to social licence considerations.

The ECJ has reacted to this gap in the famous decision *Cassis de Dijon*[11] by developing 'mandatory requirements' that a government can invoke to justify restrictions on the free movement of goods when these restrictions cannot be justified under Article 36 and when the restrictions have no discriminatory character; that is, when these restrictions do not distinguish directly or indirectly (e.g. disparate impact) between domestic and imported goods and when the restrictions put in place to defend such mandatory requirements are proportional. The ECJ mentioned in particular the 'effectiveness of fiscal supervision, the protection of public health, the fairness of commercial transactions and the defence of the consumer'.[12]

The list of potential mandatory requirements open to a state government to restrict the right of free flow of goods is open. By definition, they are justifications for existing domestic legislation that can hinder in some way the free flow of goods or (to put it more concretely) some requirement of domestic law that limits the importation and marketing of a product. Whenever such requirements have become the subject of EU regulation, usually by way of a directive to be implemented into the domestic law, state mandatory requirements are no longer relevant. The creation of a European Directive will lead to uniform standards from which state derogation is in principle not possible.[13]

Mandatory requirements, where applicable, are therefore public-interest objectives within a state that can be recognised on the EU level and will then take precedence over the free movement of goods[14] and similarly over the other market freedoms, depending on the context. They are legally acceptable limits to the social licence of industry and the citizen. Implicitly, they reflect a negotiated and supervised sense of what is reasonable and necessary in the public interest.

Cassis-de-Dijon had two important effects besides the creation of the open list of mandatory requirements. Firstly, the Court introduced the 'principle of proportionality' for assessing the validity of alleged mandatory requirements. 'Proportionality' means that the restrictive measure must be able to achieve the protection of the public interest it is meant to protect and that there is no alternative measure, which could effectively afford the same protection with less impact on the market freedom.[15] Consumer protection, for example, can be afforded by a complete product ban or by informing a consumer on the label and, where that is possible, a product ban would be disproportionate and therefore illegal.[16] This issue of proportionality contains echoes of political debates in other jurisdictions about the extent to which regulatory controls are 'overkill', and is a principle that perhaps could be more widely used in other jurisdictions, as part of processes of regulatory review.

Secondly, the Court's decision in effect stipulated a concept of 'mutual recognition'. By doing so, it revolutionised market integration. Prior to the Court's decision, a purely legislative approach was taken to aligning the laws of different jurisdictions. By way of European directives, a special form of legislation could be used to require that the implementing legislation in the member states about particular matters were to be *harmonised* – that is, streamlined and made identical – thus removing trade barriers. 'Mutual recognition', on the other hand, meant that any product legally produced and marketed in one member state could legally be sold anywhere in the Community. This *country-of-origin-principle* requires as its foundation a belief that by and large product and safety standards in the member states are similar enough to be acceptable across the EU with Article 36 TFEU and the mandatory requirements jurisprudence of the ECJ being sufficient to deal with whatever might fall outside this scope. After 'mutual recognition' there was no longer a necessity for a plethora of European legislation to harmonise the various domestic laws. The effect has been to enlarge the licence of producers to operate across jurisdictions, through their compliance with the laws in their home jurisdiction.

It is perhaps not surprising that the principle of 'mutual recognition' evoked concerns and even fears of defective, unsafe or unhealthy products or foods flooding the countries. After all, one country's specialty can be another countries health hazard as, for example, in the case of raw milk cheese.[17] This had direct impacts on the public discourse about these matters in the broadest sense. It also has had (and is likely to increasingly have) significant impacts on the freedom of commerce within states, because if a product is legally available within a jurisdiction then the arguments against manufacturing it within that jurisdiction become much harder to support. Extension of the scope of the licence of commerce is a likely result of mutual recognition.

'Mutual recognition' did not make all EU legislation unnecessary, but made it possible for the EU to concentrate on *vertical* legislation: that is, legislation across a range of products.[18] The debate on legislation dealing with admissible food additives or the admissibility of genetically modified foods directly touched on social licence issues because the acceptance for such supra-national legislation in the various member states was at stake. It is in such debates where it is conceivable that member states' governments might quietly support measures on the European level and at the same time take a much more reserved stand at home.[19]

Article 36 TFEU and mandatory requirements as instruments for social licence considerations

As discussed earlier, the Article 36 exceptions and the mandatory requirements are public-interest objectives, which the member states can put forward to justify common market restrictions. Coupled with the principle of 'mutual recognition', the mandatory requirements are domestic public interest considerations that can be held against the importation of a product from another member state, even if that product is lawfully produced and marketable there. From a policy perspective, the mandatory requirements are decentralisation instruments. 'Mutual recognition' in effect elevates one country's standard to the EU standard in that area. Mandatory requirements allow the receiving member states to defend their home standards against the state of origin under certain, but not unlimited, circumstances.

It is obvious that acceptance issues play an important role. In the light of the fact that the dogmatic structure of the common market in EU law has become entrenched and has led to strict supervision of the common market principles by the stakeholders under the rule of law, the mandatory requirements are an important element to uphold national peculiarities and specific elements of the national legal system. It is not coincidental that this interacting system of 'mutual recognition' and mandatory requirements is often referred to as the 'rule of reason'.[20]

This reflects core social licence considerations. The law itself (more so in intertwined multi-layer federal systems with strong elements of *foreign* (in the sense of non-domestic)) regulation is based on social licence considerations to secure acceptance of the legal order in the various parts of the jurisdiction.

The 'principle of proportionality' as an instrument for social licence considerations

Closely related to the mandatory requirements as counterweights of the principle of 'mutual recognition' is the 'principle of proportionality'.[21] Mandatory requirements can only be invoked if the restrictions caused by the protective measures to the common market are proportional. Proportionality is a two-pronged test resting on the capability of the measure in question to achieve the intended protection of the public interest in question and the necessity of this measure; that is, the question of whether there is a less intrusive measure that could ensure the protection of the public interest with equal effectiveness, but less intrusiveness, to the common market.

The necessity test requires that the action taken in protection of some legal interest be compared with other possible actions that could be taken to protect that same interest. The determination of whether some alternative action has equally protective qualities as the one chosen, or to be chosen, depends on an evaluation of the risks involved. The higher the potential damage if the protective goal is not achieved the more difficult will it be to show that an alternative method of protection of lesser intensity is equally effective, and hence to be preferred. The considerations involved in this exercise can be, and are often, driven by social licence considerations. In effect, the ECJ acts as a mechanism of legislative review against these standards.

Two examples illustrate this. In the famous beer case referred to earlier,[22] the ECJ came to the conclusion that the German prohibition to market beer not brewed under the purity standards defined in the law was disproportionate. Labelling could have achieved the same level of protection. That is undoubtedly true, but the finding rests on the fact that the risk incurred by a consumer drinking an impure beer (i.e. one brewed using rice instead of barley) is minute or even non-existent. Should a consumer not read the label, he or she will have nothing to fear other than perhaps a disappointing experience. Contrast this to the bovine spongiform encephalopathy (BSE) crises in Britain and the ensuing total ban on the sale of British beef in the EU[23] and elsewhere. There was never much evidence of BSE being transmittable by eating meat as such, as opposed to other parts of the animal (especially nervous system tissue). However, much was, and is, unknown[24] but it is inconceivable that despite the absence of an established nexus between the meat and the (new variant) Creutzfeldt–Jacobs disease, a regulator could have proposed to deal with this issue merely by labelling the beef and leaving it to the consumer to make a decision. Such an approach would not have violated the community's expectation of the role of government in limiting the social licence to operate; that is, it would not have gained acceptance in the population and thereby it would have damaged trust in the regulatory bodies, and arguably failed in one of the principle roles of democracy.

Derogating domestic legislation under Article 114 TFEU as an instrument for social licence considerations

The EU is built on the principle of enumerated powers; that is, the EU can only exercise legislative powers when that exercise has a concrete foundation in the TFEU. One important provision in the TFEU regarding common market legislation is Article 114 TFEU.[25] Under the title 'Approximation of Law' the provision gives the EU the right to legislate in the form of directives with regard to the 'establishment and functioning of the internal market' (Article 114.1 TFEU). The provision is a fallback clause for common market related issues to be legislated on the EU level if no legislative power can be found in other provisions of the treaty. Of interest in considering social licence are those parts of the provision allowing member states under certain conditions to derogate from the EU legislation they are in principle bound to transform into national law.

The starting point is Article 114.3, which obligates the European Commission as the sole initiator of European legislation[26] to:

> '*... take as a base a high level of protection' when submitting legislative proposals concerning health, safety, environmental protection and consumer protection in the common market.*'

Such legislation is passed in the form of a directive, which the member states must transform into their domestic law. Article 114.4–6 contain procedural provisions under which the European Commission can approve or reject a member state's request to maintain or introduce into its domestic law provisions derogating from such a directive. Such a request is permissible in order to protect the public interests spelled out in Article 36 and relating to the protection of the environment or the working environment. Thus it is possible for a state, where it has adequate public good reasons, to restrict the licence of enterprises within that state to a degree that may not be legitimate in other states of the Union.

Thus there are two ways in which the social licence of business may be legitimately restricted, even in the face of the over-arching rights associated with freedom to operate across the Union. Under Article 36, justifications for maintaining trade barriers (and their brethren, the mandatory requirements) are considered after the fact when a market participant challenges domestic restriction on the grounds that the restrictions on their activities imposed by a state violate the common market rights. Article 114 TFEU allows consideration of such restrictions before the fact to safeguard national public interest. The Commission is also called upon in Article 114.8 to examine the EU measure and suggest modifications, if deemed necessary. The principle is the same, the EU chooses not to steamroll over national concerns in areas of sensitivity. This is even despite the fact that these concerns could have been raised by the member state in the legislative process and, where this may have occurred, despite the fact that the majorities in the legislative process were apparently not convinced of the significance of the concern, or thought it was better to have the legislation in force in most members states even if one or two members states will subsequently derogate.

The beef-hormone dispute between the EU and the USA

A brief excursion into World Trade Organisation (WTO) law yields a similar example of how social licence considerations might interact with trade measures. It also illustrates the role of supra-national rules in seeking to call into question (and requiring objective justification for) state imposition of restrictions on the social licence to operate of businesses within their jurisdictions. The interaction between science and politics as a basis for restricting the licence of industry to operate is also illustrated.

The beef-hormone dispute was about the use of growth hormones in beef production, which was allowed in the USA and Canada but not in the EU. This led to a dispute under the WTO dispute settlement system[27] because the USA and Canada alleged a violation of the 'Agreement on the Application of Sanitary and Phytosanitary Measures' (SPS-Agreement[28]). This requires in Article 5 that member states undertake a science-based risk assessment of their sanitary or phytosanitary measures.[29] However, there is no scientific proof that the growth hormones could have any material adverse effect on the consumer.[30] Both the Panel and the Appellate Body therefore found in favour of the USA. The problem was not so much caused by protectionist sentiments in the EU, although there was certainly relevant but negligible side effect.[31] The main problem was a very strong consumer sentiment against growth hormones in most member states. For the EU to accept the legal obligations under the WTO rules without putting up a fight to protect the European consumer would have come at a very heavy political cost at home.

It can safely be concluded that this whole dispute owes its existence not so much to differences in legal opinion on the interpretation of Article 5 of the SPS-Agreement but much more to social licence considerations that forced the EU to act in order to defend community perceptions of the boundary that they wanted to see around the commercial licence to operate. Government needed to be seen to be emphatically reflecting strongly held and extremely negative attitudes to growth hormones in beef production among their constituencies. This example also re-emphasises a point made elsewhere in this book that decisions about the social licence to operate are not governed exclusively by the dictates of law or of science: political response and the operations of the democratic process are equally involved.

Conclusion

The concept of social licence is usually used to describe the freedom that is given to commerce to exercise its economic interests free of interference. This concept includes both legal licence and action that takes place outside the limited role of the law in protecting the common good. This chapter has explored the role of social licence issues in the existence, purpose and functioning of certain legal constructs that go beyond national boundaries. It goes without saying that in a democratic society one should always (or at least often) be able to trace the existence of a legal norm to considerations that could also play a role in a social licence context.

The examples provided in this chapter go beyond the general political consideration of the pursuit of the public good when legislating or regulating within a state. They consider the imperative to maximise the licence of industry in the interests of economic efficiency and the related gains from competition across national boundaries, and the pressures of democratic action within these boundaries to restrict this competition in favour of other interests or concerns. They show the tension between internationally agreed rules to defend the maximum freedom of commerce to operate unhindered, and the desire within states to restrict that social licence in response to particular localised concerns about the public interest. In this dynamic, we can see some complicating aspects of the operation of democracy within states, as well as the effects of international relationships that impose obligations upon states.

In the EU, two aspects help explain this dynamic. Firstly there is the EU's peculiar federal structure. It differs from other federal structures in that the alignment of political interests between the various constituent entities is dissimilar to classical federations, where these interests are either segregated by allocation to either the federal centre or the constituent entities or where these interests are aligned in a much more unitary and less antagonistic way, as is the case in the EU.

Secondly there is the so-called democratic deficit in the EU. Owing to the nature of the EU, democratic legitimisation cannot be achieved in the same way as in the national democracies of the classical nation state.[32] One compensating mechanism for this deficit is the principle of subsidiarity,[33] which aims to devolve as much power as possible to the lowest governmental level. The Article 36 TFEU and mandatory requirement exceptions to the free movement of goods, and similarly to all market freedoms, are in essence based on subsidiarity considerations and the underlying idea bears similarities to the social licence concept.

The chapter also points to the fact that the political/legal dynamic in which private interests are often subordinated to public interests, and where the boundaries of interest created through the law become renegotiated occurs not only at the level of the individual citizen within a state. Social licence debates at an international level demonstrate similar characteristics. What these debates do have at an international level that is lacking at a local level is a concentrated effort to define the conditions under which a state's restriction of an individual's freedom to operate can be considered legitimate, and the existence of specific legally applicable means to require reasonableness and minimal restriction, even while allowing democratic processes to remain paramount in determining the extent of economic freedom that will be evident within a state.

Endnotes

1 See, for example, the various contributions in Issues for Debate in Corporate Social Responsibility – Selection from CQ Researcher, 2009; see also N. Gunningham, Voluntary Approaches to Environmental Protection: Lessons from the Mining and Forestry Sectors, in: OECD Global Forum on International Investment, Foreign Direct Investment and the Environment – Lessons from the Mining Sector, 2002, p. 157 at 160 et seq.

2 Price T (2007) Corporate Social Responsibility – Is Good Citizenship Good for the Bottom Line? CQ Researcher, Congressional Quarterly Inc, Washington DC. see supra endnote p. 1.

3 ECJ, Case 6/64, 15.7.1964, [1964] ECR 585; see also for more background Chalmers/Davies/Monti, European Union Law, 2nd ed. 2010, Chapter 5 (p. 184 rt seq.).

4 Article 17.2 TEU: '2. Union legislative acts may only be adopted on the basis of a Commission proposal, except where the Treaties provide otherwise. Other acts shall be adopted on the basis of a Commission proposal where the Treaties so provide.' See also Article 289.1 TFEU: '1. The ordinary legislative procedure shall consist in the joint adoption by the European Parliament and the Council of a regulation, directive or decision on a proposal from the Commission. This procedure is defined in Article 294.'

5 It is not easy and highly disputed in state theory what the characteristics of a state are in contrast to an international organisation. The starting points are the three elements of territory, population and effective government. At least the latter two are problematic to begin with and of little help for the distinction between state and international organisation. One could 'measure' the 'amount' of legislative and executive powers vested in the body and compare that with the traditional nation state. Internal (police) and external security, for example, are traditionally powers held by states, and areas where the EU is still not very developed. But, even in these fields, the EU is beginning to act with a nucleus of a police (Europol, <http://www.europol.europa.eu/> already existing and several EU military missions in place (three as of July 2010, see <http://www.consilium.europa.eu/showpage.aspx?id=268&lang=EN)>. Taxation is another 'weak spot', with the EU only regulating elements of the VAT. On the other hand, it is persistently reported that about 80% of domestic legislation is in fact implementing legislation dictated by the EU. Whereas this number is probably exaggerated, and very hard to determine and prove, it does show that the EU has become a formidable legislator. See <http://www.mzes.uni-mannheim.de/publications/wp/wp-118.pdf>. The decisive element for the EU not being a state could be seen in the fact that no EU-organ is involved in the creation and amendment of the EU's constitutional structure: the founding treaties. This process is exclusively in the hands of the member states (Article 48 TEU) as the 'masters of the Treaties'. However, there are exceptions here as well. The accession of new member states by accession treaty requires the assent of the European Parliament (Article 49 TEU) and Article 48.6–7 now contains a new simplified revision procedure involving EU-institutions.

6 See Treaty of Maastricht of 1992, which also founded the European Union. With the coming into force of the reform Treaty of Lisbon on 1/12/2009 the former European

[Economic] Community has become the European Union. The European Coal and Steel Community had already been integrated into the European Community upon expiry of its founding Treaty of Paris of 1952 in 2002. The third of the original three organisation established in 1957 by the Treaties of Rome, the European Atomic Energy Community continues to exist as a separate entity but a constituent part of the EU. See <http://www.eur-lex.europa.eu/en/treaties/index.htm>.

7 Treaty on the Functioning of the European Union, see <http://www.eur-lex.europa.eu/en/treaties/index.htm>.

8 Citizenship of the European Union is conferred by Article 9 TEU: 'Every national of a Member State shall be a citizen of the Union. Citizenship of the Union shall be additional to national citizenship and shall not replace it.'

9 ECJ, Case 8/74, 11.7.1974, 1974 ECR 837, Dassonville, <http://www.eur-lex.europa.eu/LexUriServ/LexUriServ.do?uri=CELEX:61974J0008:EN:HTML>.

10 The concept was developed in ECJ, Joined Cases 267, 268/91, 24.11.1993, [1993] ECR I-6907, Keck and Mithouard, <http://www.eur-lex.europa.eu/LexUriServ/LexUriServ.do?uri=CELEX:61991J0267:EN:HTML>. At issue in this case was the prohibition and criminal sanctioning under French law for supermarkets to sell an unaltered product below purchasing price (resale at a loss). The Court did not consider this domestic prohibition to have any effect on intra Community trade as it applied to any and all products sold, and was not product based but a regulation pertaining to the sale as such (selling arrangement). The concept of selling arrangements is not without problem because the distinction can be difficult at times. See, for example, A Trifonidou, 'Was Keck a Half-Baked Solution After All?' (2007) 34 Legal Issues of Economic Integration (LIEI) 167.

11 ECJ, Case 120/78, 20.2.1979, [1979] ECR 649, Rewe-Zentral AG v Bundesmonopolverwaltung für Branntwein, <http://www.eur-lex.europa.eu/LexUriServ/LexUriServ.do?uri=CELEX:61978J0120:EN:HTML>. The case concerned a provision in the German law, which prohibited the marketing of fruit liqueurs with an alcohol under 25% by volume. Cassis de Dijon is a French blackcurrant liqueur with typically only 15 to 20% alcohol by volume and its importation for sale by the German supermarket chain REWE was therefore prohibited by German authorities. The German government justified this nonsense with public health protection needs alleging that low alcohol content products (i.e. products populating the range between wines on the one hand and the 25% plus range of spirits) 'may more easily induce a tolerance towards alcohol than more highly alcoholic beverages', id. at para. 10.

12 Id. para. 8.

13 Article 114.4-10 TFEU contains a legal basis for the derogation from harmonised law by member states on grounds that in the absence of harmonised legislation would qualify as mandatory requirements.

14 ECJ, Case C-123/00, 5.4.2001, [2001] ECR I-2795, para. 18, Criminal Proceedings Against Bellamy and English Shop Wholesale, <http://www.eur-lex.europa.eu/LexUriServ/LexUriServ.do?uri=CELEX:62000J0123:EN:HTML>.

15 The significance of the principle of proportionality in some European legal systems, most prominently Germany, and in the legal system of the EU and the European Convention on Human Rights (ECHR), cannot be overestimated. Via the ECHR, the principle has also had a significant impact on the law in the United Kingdom. See A Baker, Proportionality under the UK Human Rights Act (announced for February 2011 by Hart Publishing, see <http://www.hartpub.co.uk/books/details.asp?isbn=9781841137438>. For an overview over the principle's history and its application in a number of countries, see E T Sullivan/ R S Frase, Proportionality Principles in American Law – Controlling Excessive Government Actions, 2008, p. 15 *et seq.*

16 The German beer case is an illustrative example. Under law, only beers were marketable in Germany that were in conformity with the centuries old beer-purity law under which beer could only be made from barley, hops and water. Accordingly, beer brewed using rice or beer with other additives could not be sold in Germany. The ECJ struck down this prohibition stipulating that, whereas consumer protection can be a mandatory requirement, the complete ban of such 'impure' beers is disproportionate as labelling the beer as brewed in accordance with that standard will inform the consumer sufficiently. See ECJ, Case 178/84, 12.3.1987, [1987] ECR 1227, Commission/Germany, <http://www.eur-lex.europa.eu/LexUriServ/LexUriServ.do?uri=CELEX:61984J0178:EN:HTML>.

17 See Council Directive 92/46/EEC of 16 June 1992 laying down the health rules for the production and placing on the market of raw milk, heat-treated milk and milk-based products, OJ L 268/1 (1992), http://www.eur-lex.europa.eu/LexUriServ/LexUriServ.do?uri=CELEX:31992L0046:EN:HTML>

18 See, for example, Council Directive 89/107/EEC of 21 December 1988 on the approximation of the laws of the member states concerning food additives for use in foodstuffs intended for human consumption, OJ L 40/27 (1989), <http://www.eur-lex.europa.eu/LexUriServ/LexUriServ.do?uri=CELEX:31989L0107:EN:HTML>. This Directive contains a list of substances the use of which is authorised to the exclusion of all others, a list of foodstuffs to which these substances may be added and the conditions under which they may be added and restrictions that may be imposed in respect of technological purposes and rules concerning substances used as solvents including purity criteria where necessary; Regulation (EC) No 1829/2003 of the European Parliament and of the Council of 22 September 2003 on genetically modified food and feed, OJ L 268/1 (2003), <http://www.eur-lex.europa.eu/LexUriServ/LexUriServ.do?uri=CELEX:32003R1829:EN:HTML>.

19 See also infra Derogating Domestic Legislation under Article 114 TFEU as an Instrument for Social Licence considerations.

20 See, for example, G Moens/J Trone, Commercial Law of the European Union, 2010, p. 55 *et seq*.

21 See supra footnote 15.

22 See supra footnote 16.

23 For an overview on EU BSE legal acts, see <http://www.ec.europa.eu/food/fs/bse/bse19_en.html>.

24 See <http://www.ninds.nih.gov/disorders/cjd/detail_cjd.htm>.

25 See Articles 2–4 TFEU (Treaty on the Functioning of the European Union) and Article 5 TEU, supra footnote 7.

26 See supra footnote 4.

27 For a comprehensive list of relevant WTO-documents relating to this dispute, including the Panel decision and the report of the Appellate Body see <http://www.trade.ec.europa.eu/wtodispute/show.cfm?id=186&code=2#_wto-documents>.

28 <http://www.wto.org/english/docs_e/legal_e/15sps_01_e.htm>.

29 On the dispute and the broader context, see S Pardo Quintillán (1999) 'Free trade, public health protection and consumer information in the European and WTO Context' 33 *Journal of World Trade* 147; W Th Douma, 'The beef hormones dispute and the use of National Standards under WTO Law' 8 *European Environmental Law Review* (1999) 137; I A Sien (2007)'Beefing up the hormones dispute' 95 *The Georgetown Law Journal* 565.

30 Even the EU's experts could only provide the scantest of evidence in a way which would probably render most processed foods into the risk category, see <http://trade.ec.europa.eu/doclib/html/114727.htm>.

31 The beef would have been difficult to market in the first place. However, the final resolution of the conflict shows that the US managed to use the case to leverage their access to the European Union market for hormone free beef, see Memorandum on Beef Hormones Dispute with the United States of 14/9/2009, <http://trade.ec.europa.eu/doclib/html/145886.htm>.

32 The European Parliament can only represent the various populations of the member states in a very disproportional way by discriminating against the larger member states and privileging the smaller ones. One German member of the European Parliament represents roughly 800 000 people whereas one deputy from Luxembourg or Malta only represents about 60 000 people. Proportional representation would wipe whole countries off the map of parliamentary representation or require a Parliament more akin to the Chinese National People's Congress (and still marginalise the smaller member states). On the democratic deficit, see A Føllesdal (2006)'Why there is a Democratic Deficit in the EU' 44 *Journal of Common Market Studies* 533; S C Sieberson (2008) 'The Treaty of Lisbon and its Impact on the European Union's Democratic Deficit' 14 *The Columbia Journal of European Law* 445.

33 See for example R Schütze (2009) 'Subsidiarity after Lisbon' 68 *The Cambridge Law Journal'* 525.

Reference

Visser W, Matten D, Pohl M and Tolhurst N (Eds) (2007) *The A to Z of Corporate Social Responsibility.* Wiley Press, UK.

12

The state of social impact indicators: measurement without meaning?

Mark Shepheard and Paul Martin

> *'... Positive community functioning relies on the underlying beliefs people hold about obligation, reciprocity and philanthropy, on the prevalence in the community of attitudes such as trust, in other people and in community infrastructures, and on the extent to which individuals and groups participate in the community. Such factors can be difficult to measure but are important precursors to wellbeing at a societal level ...'*
>
> *(Australian Bureau of Statistics 2001)*

Introduction

The *social licence* questions for irrigators can be summarised as: 'how do we demonstrate that we are maximising the social gains, minimising the social costs, and reducing the social risks associated with the private use of community assets such as water?' The issues of social licence are relatively new. As a result, the approach to answering this question is as yet immature (Martin *et al.* 2007). It can be expected that there will be confusion, competing models and a complex process of evolution before the paradigm is settled and industry and the community has an accepted efficient framework (Christen *et al.* 2006; Shepheard and Martin 2009). Nowhere is this more evident than in the evolution of triple bottom line reporting. This chapter looks at recent approaches to, and proposals for, social impact indicators and suggests some directions for social impact reporting for water enterprises.

The opening quote indicates that social performance reflects the contributions made by the private organisation to social systems (Australian Bureau of Statistics 2001). In any open system such as a society, there is a plethora of interconnections and impacts in varying degrees of proximity to any action. Cause, effect and responsibility are often unclear. Specific methods are needed to define the things managers can be expected to be accountable for (Shepheard and Martin 2008). Without this, the manager faces uncertainty and boundary-less demands on resources. Once it is clear what accountabilities the manager is expected and prepared to undertake, it becomes possible to find indicators against which performance can be reported.

As yet there is no accepted mechanism for defining the boundaries of social responsibility of an enterprise. This results in the absence of logical mechanisms for determining which measures are likely to be useful and efficient (Shepheard and Martin 2008). The following review of the state of social impact indicators (focusing on irrigation in Australia) illustrates

Table 12.1: GRI social performance relevant to water management in Victoria (based on the GRI 2002 guidelines)

GRI social performance aspect and category	Measures of relevance to the Victorian Water Industry
Labour practices aspect	
Employment and decent work	Staff numbers by function; percentage of employees who are part time; percentage of employees not on permanent contract; percentage of labour requirement met by outsourcing / non employees; employee turnover as percentage of total and by type of position; percentage change in total workforce size (FTE equivalent); staff satisfaction surveys.
Industrial relations	Percentage of employees represented by independent trade unions or associations, or percentage of employees covered by collective bargaining; policy and procedure involving information, consultation, and negotiation with employees regarding changes in the organisations operations; provision for formal worker representation in decision making or management including corporate governance.
Health and safety	Number of reportable incidents; number of worker's compensation claims; number of injuries resulting in days lost (including outsourced work); days lost per employee (including contractors).
Training and education	Workforce participation in formal training; average hours of training undertaken by employee category; programs to support the development of employees; policy related to skills management and life-long learning
Diversity and opportunity	Existence of positive or affirmative action policies or programs; board composition; proportion of men and women in senior management.
Human rights aspect	
Strategy and management	Membership of Indigenous people on consultative committees; description of consultations/engagements with Indigenous communities; listing of key personnel and discussion of human rights issues in the context of the organisations operations.
Indigenous rights	Policy and management principles regarding Indigenous people.
General	Policy and procedures to evaluate and examine human rights performance, including monitoring systems and results.
Society aspect	
Customer health and safety	Health compliance reporting; number and type of major incidents and/or breaches of regulation concerning public health; report on number and nature of complaints upheld by independent regulator, e.g. ombudsman
Advertising	Advertising and promotions policy, procedures/management systems; compliance in relation to standards of social and environmental responsibility.
Respect for privacy	Policy, procedures/management systems and compliance mechanisms for customer privacy; number of substantiated complaints regarding breaches of customer privacy
Competition and pricing	Mechanism for establishing pricing; changes in prices over time
Corporate citizenship	Involvement and/or contributions to projects with value to the greater community; awards received relevant to performance.
Community	Policy, procedures/management systems and compliance mechanisms related to customer satisfaction; policies and procedures for identifying and engaging in dialogue with community stakeholders in areas affected by the organisations activity

this problem. The review synthesised the social impact indicators used by a range of irrigation authorities (see Table 12.3), and studies dealing with social impact indicators in natural resource management (identified in Table 12.7).

Indicators currently used

The Global Reporting Initiative (GRI) provides a widely accepted approach to sustainability reporting (Global Reporting Initiative 2002; Global Reporting Initiative 2006), which the Australian water industry has used (VicWater 2002). The VicWater (2002) report reviewing the use of GRI in the Australian water industry identified three aspects of social performance from the GRI guidelines with relevant indicators for the water industry. These are 'labour practices', 'human rights' and 'society' (Table 12.1). The revised GRI 'G3' guidelines (Global Reporting Initiative 2006) provide additional social performance indicators that may be relevant to the business of a rural water providers (Table 12.2). Shepheard *et al.* (2006) also provide a review of the GRI application to irrigation water providers in Australia.

Performance reports from six rural water providers show that social performance is reported across the categories of community, customers, staff and corporate governance (Table 12.3). There are a variety of indicators used, but these measures do not reflect a shared understanding of the nature of social accountability and focus substantially on matters that are internal (staff relations, asset use) rather than impacts upon society. It is possible to infer a selection of measures on the basis of data availability rather, than on the basis of an objective assessment of the social responsibility of the corporation. This is understandable given the complexity of this enquiry.

What might be sensible indicators in practice?

The concern for measurement practicality is of course sensible. Management of a corporation is principally charged with the management of resources for particular purposes in the interests of its owners, taking into account the interests of staff and customers, within the boundaries set by law. Investing substantial funds on measures that do not align with these obligations can be legitimately criticised.

Any discussion of what indicators might be efficient for water managers to adopt ought be subject to the caveat that discussion of measures before one is clear about accountability and actionability is putting the cart before the horse. Other chapters in this book have dealt more directly with the 'why' of social accountability, while we confine ourselves to the 'what' and 'how' of reporting performance. Such a discussion provides an idea about 'means', even if the 'ends' remain undefined.

Table 12.2: Additional indicators of social performance from the 'G3' edition of GRI guidelines (2006)

GRI social performance aspect and category	Measures of relevance to the water supply industry
Community	Nature, scope, and effectiveness of any programs and practices that assess and manage the impacts of operations on communities.
Corruption	Percentage and total number of business units analysed for risks related to corruption; percentage of employees trained in the organisations anti-corruption policies and procedures; actions taken in response to incidences of corruption
Public policy	Public policy positions and participation in public policy development and lobbying

Table 12.3: Indicators of social performance for rural water providers

Rural water provider	Community	Customers	Staff	Corporate governance
Goulburn-Murray Water Victoria (Goulburn–Murray Water 2006)	• Transfers in water entitlement • Volume of water traded • Action taken to meet community recreation, tourism and economic development needs associated with water bodies. • Number, location, and purpose of stakeholder reference groups, community-panels, and public meetings. • Community events supported • R & D supported	• Customer satisfaction with water supply service • Type of services and number of properties served • Number of complaints registered in complaints management system • Water order delivery performance • Translation service • Number of water services committee meetings and member attendance • Type of activity undertaken by committees • Measures taken to ease customer financial hardship	• Total number of employees and gender percentages • Staff satisfaction survey results • Number of training days, type of training and number of staff involved • OHS key indicators • Lost time injury rates reported • Number and type of women's development forums held • IT projects implemented and purpose in improving ways of working	• Mission and values statement made • Objectives and highlights listed with performance indicators and result • Challenges Identified • Board membership and experience • Board structure and practices • Directors attendance at meetings
Murrumbidgee Irrigation New South Wales (Murrumbidgee Irrigation 2006)	• Number of youth supported, type of development need met and total amount invested • Number and type of community events supported • Group progress through registered training • No of work experience students and area of work	• Bulk water performance (ML) • Sources of water • Cumulative annual water sales compared with last 3 years • Percentage of water sales per crop type • Savings initiatives and percentage allocation benefit • Progress report on integrated horticulture supply and Barren Box Storage work • Progress against MIA Envirowise (LWMP) targets	• Employee numbers • Number of injuries / time lost • Number of structures automated • Time lost connected with major works • Implementation progress of performance management system • Type of staff development opportunities supported	• Company code of conduct • Governance statement made • Directors responsibilities listed and interests declared • Directors attendance at meetings • Management statement on finances and income

Rural water provider	Community	Customers	Staff	Corporate governance
Murray Irrigation Limited New South Wales (Murray Irrigation Limited 2006)	• Number of industry alliances formed and type networks • Number and type of community events sponsored • Compliance with the National Water Initiative • Stakeholders and key issues identified • External projects and partners identified	• Customer satisfaction with water supply (Australian National Committee on Irrigation and Drainage: ANCID) • Customer satisfaction with communication (ANCID) • Customer satisfaction with service (ANCID) • Provision of water information to customers • Number of talking water recipients and frequency of issue.	• Key corporate culture strategies listed • Staff turnover • Maintain OHS and QA accreditation • Reduce lost time injuries • Workplace profile given • Training and development undertaken • Injury and accident summary • Resource use and reuse summary	• Reporting boundaries identified • Number and purpose of meetings with regulatory agencies • Directors code of conduct listed • Board responsibilities listed • Board training and advice sought • Directors report • Challenges listed • Financials
SunWater Queensland (SunWater 2006)	• Public safety initiatives undertaken • Community events supported • Support for recreation at storages • Minimise risks to public through spillway adequacy assessments and comprehensive risk assessments on dams • Number and location of community water treatment facilities constructed	• Improved water trading systems • Progress on development of water sharing approaches • Achievement of service targets • Customer participation in scheme management • Communication with customers	• Learning and development systems in place • Leadership development programs • Technical training undertaken • Staff satisfaction survey • OHS performance	• Company profile – vision, mission and values • Structure • Financials • Board membership and experience • Board operations • Key compliance areas listed

Table 12.3: (Continued)

Rural water provider	Community	Customers	Staff	Corporate governance
Southern Rural Water Victoria (Southern Rural Water 2006)	• Number and type of community programs supported in the region • Focus areas for communication are listed • Number and type of partnerships established. • Type and location of community and stakeholder engagement activities is provided	• Number of customer committees and briefings held • Complaints management process detailed • Value of community service obligations listed • Progress of corporate family workshop • Diversity in committee membership • Function of licensing business forum • Details of transparent licensing process • Participation in water management planning	• Staff diversity • Number of employee consultative committee meetings held • Number of Equal Employment Opportunity officers trained • Activity of the women's network and its aims. • Training undertaken and development opportunities supported • Number of incidents and lost time associated. • OH&S development activities undertaken.	• Governance targets listed • Issues for further work identified from board self assessment • Strategic planning, management and governance arrangements documented • Values statement made • Financials
State Water New South Wales (State Water 2005)	• Number and purpose of memorandum of understanding (MoU) with regulatory organisations and stakeholder groups • Location and participants in tours and briefings for stakeholders	• Key outcomes and achievements of customer service committee meetings • Water allocation, flows, and ordering communication innovations • Community reference panel progress. • Achievements reported against customer service charter	• Staff communication activity • Opportunities for staff feedback into decision making identified • Management and strategy reforms identified and benefits to staff identified • Number of staff • Turnover rate • Sick leave taken • Lost time to injuries • Learning and development programs undertaken and amount invested	• Functions listed • Board member profiles and attendance at meetings • Committee structure of board and other statutory details of board arrangements • Staff diversity compared with salaries • CEO performance statement • Compliance report • Consultancies • Publications • R&D supported • Waste management statement • Water price details • Number of customers • Number of licences

An indicator is '... a quantity that can be measured directly and used to track changes over time with respect to an operational objective ...' (Chesson 2004). Determining indicators for bulk water providers to measure their impacts on society requires a focus on community well-being and social networks as the link between corporate strategy and the social setting. This people-focused 'grounding' of indicators, is necessary to account for local circumstances and the effect of locally unique power relations between stakeholders (Lockie *et al.* 2002).

The correlation between the indicators that are used and objective assessment of the water corporation's possible accountabilities is understandably loose. The preceding tables show that, from the constellation of relevant measures suggested by the GRI, companies consistently select a narrow band, which has little to do with measurable impacts upon community social welfare. This narrow focus on reporting what can be readily measured from internal records, rather than determining what socially significant impacts of the organisation may in part reflect a limited awareness of the types of data that can be generated that might provide more meaningful insights. Measures are selected on the basis of 'measurability' itself.

Indicators of community well-being and networks, within a strategic framework and regional setting, are found in *Signposts for Australian Agriculture* (Chesson *et al.* 2005); *Socio-economic Indicators & Protocols for the National NRM M&E Framework: Social and Institutional Foundations of NRM* (Fenton 2006);) *Social Assessment Handbook: A Guide to Methods and Approaches for Assessing the Social Sustainability of Fisheries in Australia* (Schirmer and Casey 2005), and in discussions of social capital by the Australian Bureau of Statistics (2001; 2004; 2006) and the Department of Transport and Regional Services (2005). Generic indicators may also be sourced from the GRI guidelines (Global Reporting Initiative 2006).

Social indicators are under development for networks and capacity building between regional organisations and community (Fenton 2006). These might be adapted to suit social performance of bulk water organisations, reflecting concerns for the role of the corporation in fostering:

- engagement: a shared vision and ownership at the regional level
- partnerships: underpinned with trust and confidence
- capacity: organisations with capacity to make decisions on regional issues and programs (e.g. this might relate to water management)
- recognition: the social foundations of management planning and decision making being recognised.

Corporate concern for fostering such attributes in 'their' community would reflect a positive approach to engagement and a desire to foster meaningful, long-term social relationships around issues of social licence. Our synthesis of the social impact reported by regional bulk water providers suggests that 'relationships', 'communication' and 'service' are the key social themes, but that the perspective on these issues is narrow.

Identifying relevant social considerations

'Well-being' has been defined as '... a state of health or sufficiency in all aspects of life ...' (Australian Bureau of Statistics 2004). Well-being is a notion relevant to measuring and managing social impact and natural resource stewardship. Measuring well-being involves understanding the influence of key areas of social concern and the role that transactions play (see Figure 12.1). The model in Figure 12.1 suggests a framework for water providers social performance based on the contribution that the organisation makes to the well-being through individuals within the business (the organisation's human capital), and of the community (human systems external to the organisation).

Social issues affecting well-being are likely to vary with the types of communities and water enterprise activities. For example, a water utility located in a community of high social disadvantage may find that fostering local workforce participation is a key consideration, whereas a

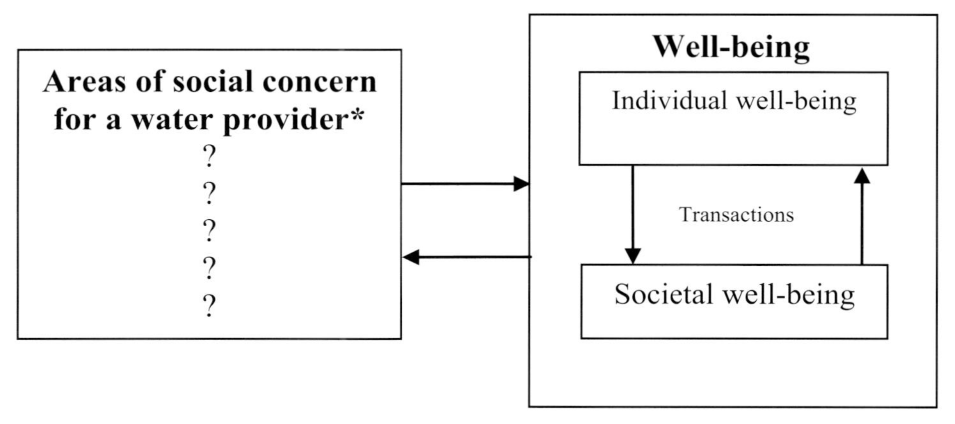

Figure 12.1: Framework for measuring well-being (adapted from ABS 2001). (*Areas of social concern assume relevance through accountability and actionability by the enterprise.)

water utility in an industrial setting with high employment may better define its social contribution through specific causes. Identifying key social issues within the host community requires a social assessment to determine the potential contribution to well-being within and beyond the organisation. A range of social data are relevant in determining what social performance issues exist (Schirmer and Casey 2005). Potential indicators that an irrigation water provider may see as relevant to the well-being of a region are illustrated in Table 12.4. Some of the data types listed in the table can be gathered within the organisation, from other organisations or from projects (Sinclair Knight Merz Pty Ltd 2003).

It has been suggested that the framework like that identified in Figure 12.1 can be a starting point for developing indicators, using the following loose process (Australian Bureau of Statistics 2001):

1. Think about the scope and definition for each area of social concern.
2. Consider the relationship of each area of social concern to well-being (within the organisation and to the external community).
3. Identify the key social issues.
4. Identify the key population groups.
5. Develop or make use of conceptual frameworks about each social concern and its issues,.
6. Identify data sources or information needs.

The focus of such an evaluation is not only the potential effects (positive or negative) upon individuals. A large part of social capital and welfare derives from interactions within networks of people. A social network is a group of people with patterns of interaction or ties between them (Scott 2000). Networks can develop shared norms, values and understanding and provide the cooperation that leads to changes in well-being (Organisation for Economic Cooperation and Development 2001). Networks are enablers of performance, by which transactions are undertaken. Networks can be interpreted using the four characteristics of qualities, structure, transactions and type (Australian Bureau of Statistics 2004). These characteristics provide the basis for measuring the performance of networks as highlighted in Table 12.5.

In this section, we have identified potential issues of individual and collective welfare. The third dimension of social obligation or responsibility is the linkage between social measures and management accountability. Responsibility can be defined in terms of the entities'

Table 12.4: Information relevant to determining relevant issues

Information type	Detailed elements
History of irrigation	• Development of current irrigation communities • Regulation and management of irrigation over time • Irrigation methods and practices
Social profile of irrigation and general communities	• Age, gender, education, income • Dependence on irrigation • Employment/unemployment • Other characteristics
Quality of life	• Life satisfaction • Work satisfaction/work conditions • Stability of/access to industry • Training and education opportunities • Physical and mental health • Income (irrigation and non-irrigation)
Social capital	• Networks • Access to services • Links to family, friends, irrigation sector, broader community
Values, attitudes and beliefs	• Of irrigation community • Of general community about irrigation
Spatial links between water resources and communities	• Examining the locations of social activities and populations in relation to the aquatic resources they use

Adapted from Schirmer and Casey 2005

accountability and the social issues where it has some commitment to action. Responsibility is both imposed by the outside world as an obligation (legal or moral) and voluntarily assumed (informally or by some form of contract or statement of commitment). This is akin to 'staking a claim' on the impacts for which the organisation is prepared to be accountable: creating a link between organisational strategy, indicators of performance and social systems.

Table 12.5: The characteristics of performance for social networks

Characteristic	Attributes of performance
Network qualities	• Norms: trust/trustworthiness, reciprocity, sense of efficacy, cooperation, acceptance of diversity, inclusiveness. • Common purpose: social participation, civic participation, community support, friendship, economic participation.
Network structure	• Size • Openness • Communication mode • Transience/mobility • Power relationships
Network transactions	• Sharing support: physical/financial assistance, emotional support, encouragement, integration into community, common action. • Sharing knowledge: skills and information, introductions. • Negotiation • Applying sanctions
Network types	• Bonding • Bridging • Linking

Determining the boundaries of the corporations imposed or assumed responsibility is an important step in determining what social indicators are to be used. It is managerially meaningless to report on things that do not fit within the corporations boundaries of responsibility. Businesses are increasingly expected to act responsibly, operating as a positive social force and working in partnership to deal with issues within society, including environmental stewardship and ecologically sustainable development (McKay 2006; Warhust 2005). Responding to these demands requires a strategic and collaborative approach to the underlying values, socio-cultural norms and perceptions that exist about the organisation, its ethical obligations, its operations, and the relevant boundaries of responsibility (Longstaff 2000; Muller and Siebenhuner 2007; Shepheard and Martin 2008). A critical point in deciding what to report upon is to make specific choices about what accountabilities and responsibilities society will expect of it and, beyond that, what additional social good interests it believes it can advance without unduly distorting its managerial obligations. Once this difficult step has been taken, it is possible to make meaningful decisions about indicators.

Research on social indicators

Research on social indicators that are relevant to water management organisations come from a range of sources (Table 12.6).

This research suggests that meaningful indicators ought to be the result of a strategic approach. That approach involves a disciplined, carefully reasoned analysis. It begins with understanding the community upon which the corporation might have either positive or negative effects. It involves careful consideration of what legal or moral or political expectations exist that create potential responsibilities upon the corporation. It also involves consideration of the potential 'low hanging fruit', where the corporation can generate positive social gains without distorting its core function.

Out of such an analysis, it is possible for a corporation to be specific about the boundaries of its social responsibilities against which it expects or is prepared to be accountable. This provides a basis of integrity in choosing on what indicators of social performance it wishes to report.

The research projects identified in Table 12.6 have documented specific impacts on society and their associated indicators of performance. These are shown in Table 12.7 as a guide to the

Table 12.6: Key projects related to social impact

Project	Social impact focus	Organisation
Signposts for Australian Agriculture	Contribution of social systems to sustainability	Bureau of Rural Sciences.
National NRM Monitoring and Evaluation Framework	Capacity for communities to respond and manage for effective NRM outcomes	National Land and Water Resources Audit. Australian Government (regional NRM)
Sustainability Challenge	Improved governance, and contribution to community well-being	Cooperative Research Centre for Irrigation Futures
Social Assessment Handbook	Planning and carrying our social assessments.	Bureau of Rural Sciences
Irrigation Management Information and Reporting System	Social impacts from irrigation on the community	Murray Darling Basin Commission

Table 12.7: Indicative social impacts and indicators

Impact	Indicators
Managing for effective NRM outcomes (Fenton 2006)	Engagement: a shared NRM vision and ownership at the regional level. Partnerships: NRM partnerships between government and regional organisations are underpinned with trust and confidence Capacity: regional bodies capacity to make decisions on NRM Recognition: governments and regional organisations recognise the importance of the social foundations of NRM.
Social systems contribution to sustainability (Chesson 2004)	Human assets, contribution to human systems
Industry contribution to human well-being (Schirmer and Casey 2005)	Indigenous well-being, local and regional well-being, national social and economic well-being
Contribute to improved well-being and quality of life of rural communities (Christen *et al.* 2006)	• Includes improved staff satisfaction, conditions of employment and equality of opportunity, with the indicators being: staff numbers and turnover, staff survey on satisfaction, staff benefits above award/industry standard, work-life balance, lost time injury rates, staff attendance rates, training expenditure as a percentages of base salaries, career development programs, gender breakdown (board, management, staff income), workforce diversity (percentage of women, ethnic diversity), affirmative action policies or programs, breakdown of workforce by employment and contract type (including work that has been contracted out), and industrial relations mechanisms. • Includes building capacity for improved farm management, with the indicators being: changes resulting from dialogue with commodity groups, measures of impact from farm financial and irrigation/ agronomic management courses. • Includes contribution to the improved well-being to those living and working on farms, with the indicators being: measures of impact from education support projects, level of satisfaction with recreation facilities and services, injury rates among farm workers, changed practices resulting from OHS training for on farm workers. • Includes contribution to improved well-being for those living in rural towns and surrounding areas, changes resulting from dialogue with authorities responsible for facilities such as education, health, employment and transport,, involvement and contribution by organisation to projects with value to the greater community.
Contribute to improved community identity, representation and participation (Christen *et al.* 2006)	• Includes increased engagement with stakeholders, with the indicators being: key activities of the company against key stakeholder groups engaged/consulted, survey of stakeholders to identify issues of concern. • Includes improved community value of cultural and heritage assets, percentage of projects that considered cultural heritage issues as ongoing part of its development.

Table 12.7: (Continued)

Impact	Indicators
Improved organisation governance (Christen *et al.* 2006)	Includes increased quality of service to customers, with the indicators being: customer satisfaction with service, infrastructure development and maintenance, security of water supply and process, efficiency of water delivery, farm surveys, research and development projects, communication strategy, awareness of water politics and associated community actions, field days, number of positive/negative media stories.
Customer satisfaction (Australian National Committee on Irrigation and Drainage 2007)	• Customer satisfaction with water supply • Customer satisfaction with communication • Customer satisfaction with service

sort of issues and indicators that are relevant for irrigation water providers. Indicators should be adopted if they can be linked to strategic planning within the organisation, maintaining the critical approach that indicators are developed as part of a strategic business management. Indicators used should also meet criteria such as the SMART filter (Alexandra and Associates Pty Ltd 2000) modified for water management organisations (Christen *et al.* 2006). SMART refers to the criteria for indicators to be: simple, measurable, accessible, relevant and timely.

Achieving a strategic approach to social performance management requires a process to connect the planning and management actions of an irrigation water provider with the relevant potential impacts. The Bureau of Rural Sciences proposes using a component tree to achieve this (Chesson 2004). In this approach, the contributions to social systems are a key heading for planning (see Figure 12.2). Component trees provide a mechanism for working through objectives from strategy to operational level, on which basis indicators can be rationally selected (Christen *et al.* 2006). Component trees can be a focus for a participatory approach (Chesson 2004), which can ensure sensitivity to local social systems (Lockie *et al.* 2002). Such an approach implicitly identifies the boundaries of responsibility that the organisation sees as relevant, reducing the risks of unbounded accountability, and therefore unbounded reporting.

The gaps

Our detailed examination of rural water suppliers provides a case study to illustrate social licence issues that go beyond this sector. Although the move to public reporting of social impacts is to be applauded, the developments are fledgling. A great deal of the decision making appears to be made with an eye to public relations and convenient measures. There is little evidence of the sort of strategic rigour that is applied to other key aspects of corporate responsibility and corporate strategy. Unless there is clarity about the boundaries of what the organisation is intending its managers to be accountable for, and what it is intended to act upon, community faith in such reporting may be lost, and the costs of attempting to do the right thing become disproportionate to the social and institutional benefits.

Rural bulk water providers particularly report on community, customers, staff and governance issues. However, a sound link between corporate strategy and social responsibility remains to be fully developed. Without this, the rationale for the selection of the communities, issues and measures is weak. Figure 12.3 suggests a conceptual framework for integration of corporate social reporting within the strategic management system. It highlights the three significant gaps, which we have identified in this case study, that inhibit making these connections.

These gaps expose water enterprises to avoidable problems. They include the risk that the corporation may invest in reporting on matters that it has no capacity to manage, or where

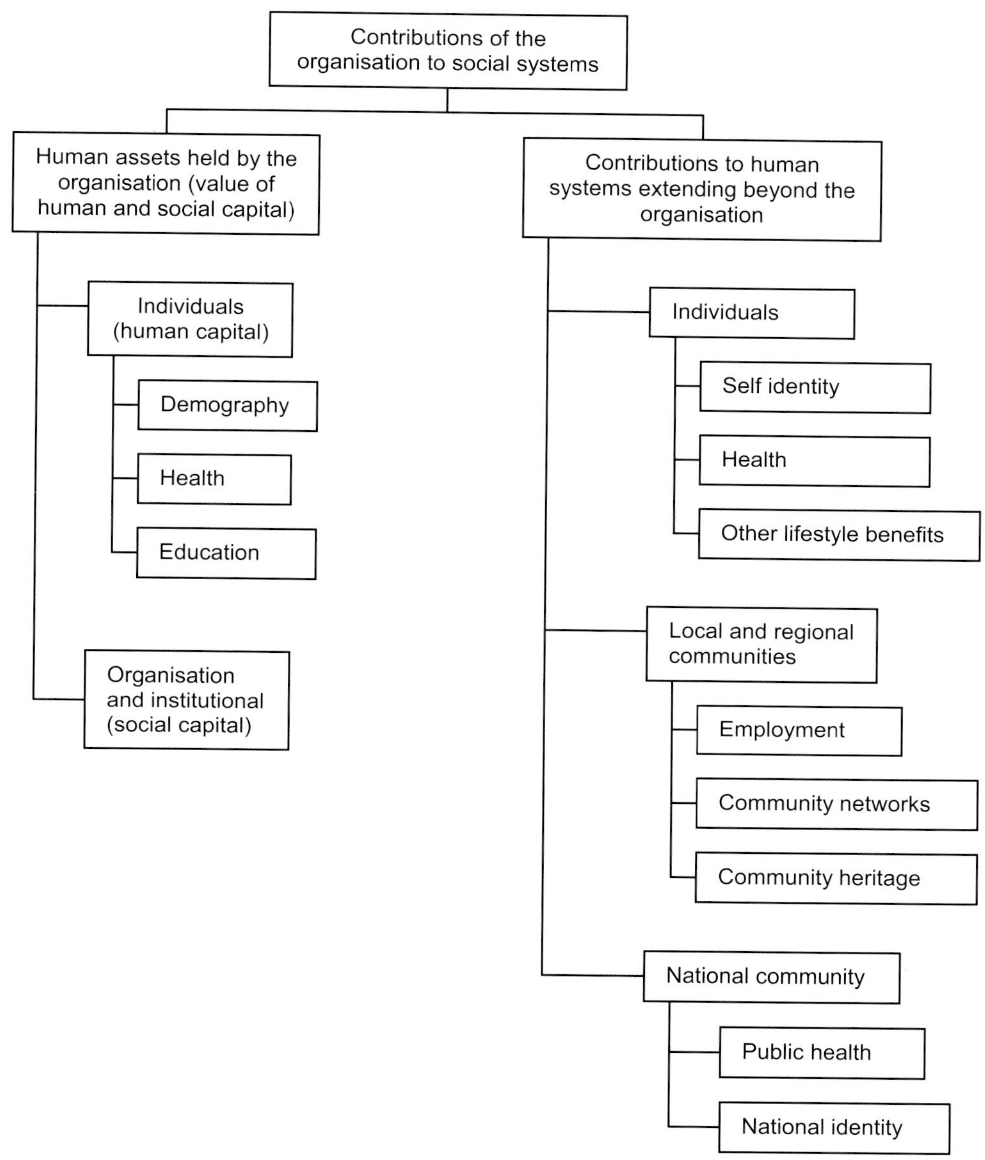

Figure 12.2: Component tree development for social systems (adapted from the Signposts for Australian Agriculture Program)

investment may not be a sensible balancing of its social accountability and its shareholder, staff and customer responsibilities. Without a rational justification for where it draws its accountability boundaries (and therefore why it measures and intervenes as it does) it may become difficult for the organisation to credibly defend the choices that it makes. Its defence of its social licence may be weakened, even as the corporation strives to 'do the right thing'.

There appears little evidence of a consensus about the boundaries of responsibility for the water corporation, or indeed any other type of natural resource user. As a result, what is being measured and how it is being measured is heterogeneous. This represents a barrier to any move towards a meaningful implementation of social performance measurement and reporting.

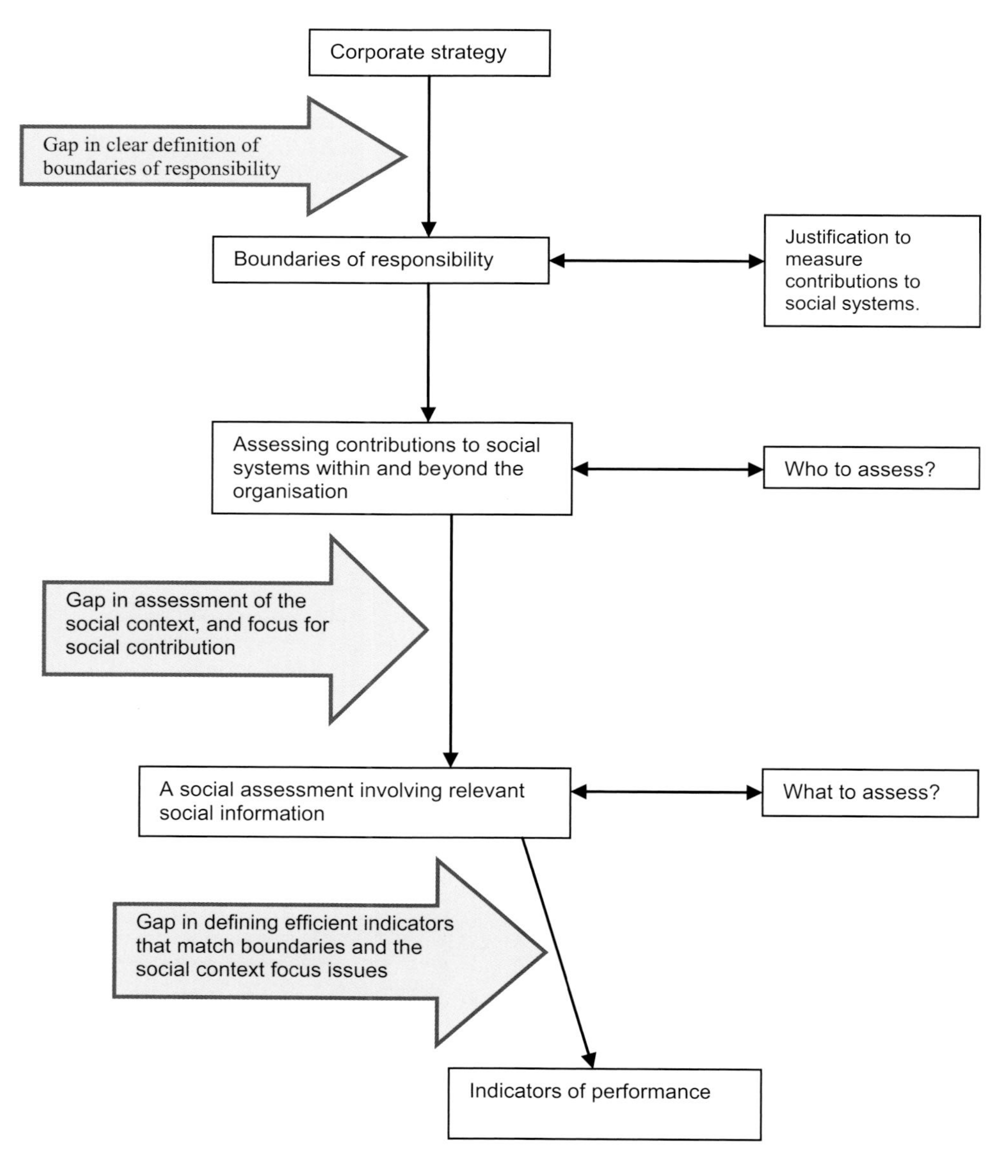

Figure 12.3: Gaps in the process of developing strategic indicators of social performance for stewardship of water by rural water providers

References and further reading

Alexandra and Associates Pty Ltd (2001) 'Sustainability indicators for north-east Victoria.' Alexandra and Associates, Eltham, Victoria.

Australian Bureau of Statistics (2001) 'Measuring wellbeing-frameworks for Australian social statistics'. ABS catalogue no. 4160.0, ABS, Canberra, <http://www.abs.gov.au>.

Australian Bureau of Statistics (2004) 'Measuring social capital – an Australian framework and indicators'. ABS Information Paper. ABS catalogue no. 1378.0, ABS, Canberra, <http://www.abs.gov.au>.

Australian Bureau of Statistics (2006) 'Measures of Australia's progress'. ABS catalogue no. 1370.0, ABS, Canberra, <http://www.abs.gov.au>.

Australian National Committee on Irrigation and Drainage (2007) 'Benchmarking data report for 2005/2006 key irrigation industry statistics and performance indicators'. ANCID Torrens, ACT.

Chesson J (2004) Measuring contributions to sustainable development – a practical approach. *Public Administration Today* December–February 2004–2005, 6–13.

Chesson J, Whitworth B and Bureau of Rural Science (2005) *Signposts for Australian Agriculture*. Australian Government, Canberra.

Christen E, Shepheard M, Jayawardane N, Davidson B, Mitchell M, Maheshwari B, Atkins D, Fairweather H, Wolfenden J and Simmons B (2006) 'The sustainability challenge: a guide to using triple bottom line reporting as a framework to promote the sustainability of rural and urban irrigation in Australia'. CRC for Irrigation Futures Technical Report No. 03-1/06. CRC for Irrigation Futures, Richmond.

Department of Transport and Regional Services (2005) 'Focus on Regions No. 4 Social Capital'. Bureau of Transport and Regional Economics Information Paper No. 55, Canberra.

Fenton M (2006) 'Socio-economic indicators & protocols for the national NRM M&E framework: social and institutional foundations of NRM'. National Land and Water Resources Audit, Canberra.

Global Reporting Initiative (2002) *Sustainability Reporting Guidelines*. The Global Reporting Initiative Amsterdam, The Netherlands.

Global Reporting Initiative (2006) *Sustainability Reporting Guidelines*. The Global Reporting Initiative Amsterdam, The Netherlands, <http://www.globalreporting.org/Reporting-Framework/G3Online>.

Goulburn Murray Water (2006) 'Annual report'. Goulbourn Murray Water, Tatura, Victoria, <http://www.g-mwater.com.au>.

Lockie S, Lawrence G, Dale A and Taylor B (2002) Capacity for change: testing a model for the inclusion of social indicators in Australia's national land and water resources audit. *Journal of Environmental Planning and Management* **45**(6), 813–826.

Longstaff S (2000) Corporate social responsibility. *City Ethics* **40, Winter**.

Martin P, Bartel RL, Sinden JA, Gunningham N and Hannam I (2007) 'Developing a good regulatory practice model for environmental regulations impacting on farmers'. Land and Water Australia, Canberra.

McKay J (2006) Issues for CEOs of water utilities with the implementation of Australia's water laws. *Journal of Contemporary Water Research and Education* **135**, 115–130.

Muller M and Siebenhuner B (2007) Policy instruments for sustainability oriented organisational learning. *Business Strategy and Environment* **16**, 232–245.

Murray Irrigation Limited (2006) 'Annual report'. Murray Irrigation Limited Deniliquin, New South Wales, <http://www.murrayirrigation.com.au>.

Murrumbidgee Irrigation (2006) 'Annual report'. Murrumbidgee Irrigation, Griffith, New South Wales, <http://www.mirrigation.com.au>.

Organisation for Economic Co-operation and Development (2001) 'The wellbeing of nations: the role of human and social capital, education and skills'. OECD Centre for Educational Research and Innovation Paris, France.

Schirmer J and Casey AM (2005) *Social Assessment Handbook: A Guide to Methods and Approaches for Assessing the Social Sustainability of Fisheries in Australia*. Fisheries Research and Development Corporation ESD Reporting and Assessment Subprogram Publication No. 7, FRDC, Deakin, ACT.

Schwartz MS and Carroll AB (2003) Corporate social responsibility: a three domain approach. *Business Ethics Quarterly* **13**(4), 503–530.

Scott J (2000) *Social Network Analysis: A Handbook*. 2nd edn. Sage Publishing, London.

Shepheard ML, Christen EW, Jayawardane NS, Davidson P, Mitchell M, Maheshwari B, Atkins D, Fairweather H, Wolfenden J and Simmons B (2006) 'The principles and benefits of triple bottom line performance reporting for the Australian irrigation sector'. Technical report No. 03-2/06. CRC for Irrigation Futures, Richmond.

Shepheard ML and Martin PV (2008) Social licence to irrigate: the boundary problem. *Social Alternatives* (special issue on Justice and Governance in Water) **27**(3), 32–39.

Shepheard ML and Martin PV (2009) The multiple meanings and practical problems with making a duty of care work for stewardship in agriculture. *Macquarie Journal of International and Comparative Environmental Law* **6**, 191–215.

Sinclair Kinght Merz Pty Ltd (2003) 'IMIRS Stage 2 – Goulburn Broken case study'. Case Study Report. Murray Darling Basin Commission, Canberra.

Southern Rural Water (2006) 'Annual report'. Southern Rural Water, Maffra, Victoria, <http://www.srw.com.au>.

State Water (2005) 'Annual report'. State Water, Dubbo, <http://www.statewater.com.au>.

SunWater (2006) 'Annual report'. Sun Water, Brisbane, <http://www.sunwater.com.au>.

VicWater (2002) 'Triple bottom line reporting guidelines'. Victorian Water Industry Association, Melbourne.

Warhurst A (2005) Future roles of business in society: the expanding boundaries of corporate responsibility and a compelling case for partnership. *Futures* **37**, 151–168.

13

The business judgement rule and voluntary reporting

Christopher Stone and Paul Martin

Some sectors of society wishing to see commerce adopt high standards of corporate citizenship may feel that those managers who do not enthusiastically embrace voluntary social or environmental reporting are trying to avoid their obligations to the broader community. This may be a naïve and unfair judgement. The job of the corporate manager is to use other people's money to meet other people's goals. With the role comes a moral and legal responsibility. A challenge for the responsible steward is to balance the private interest of the owners against the public desire for more information about corporate social performance. In this chapter, we explore the line between a narrow perspective on a manager's legal responsibility to be frugal with corporate resources, and the growing expectation that management will spend some of these resources on corporate reporting that may in itself increase pressure on the corporation to spend further resources pursuing social ends.

The legal duty to be frugal

In Australia, the duty of the director or officer (manager) is generally to act diligently in the interests of the company,[1] and within this to apply a 'fiduciary' standard to the use of their shareholder's resources.[2] This would, on the surface, suggest that the legal and moral duty of the manager is to be frugal with the use of resources and to concentrate narrowly on the economic interest of the ultimate owners of corporate capital.

As will be illustrated in this chapter, the reporting burdens on corporations that use natural resources are substantial and expensive to satisfy. Additional voluntary reporting will add to the cost and complexity, and may create new risks to the economic interests of shareholders. A responsible steward would not assume these imposts merely because others who are not shareholders would like to see things done differently.

There have been attempts in different countries to broaden the manager's social responsibility.[3] In some jurisdictions, the result has been to redefine the legal obligation of the manager to take into account the interests of stakeholders beyond those with a direct economic interest.[4] In other countries, including Australia, the dilemma of balancing the legal and social obligations of the corporation has been left in the hands of the manager, aided by court interpretations of the meaning of the statutory and common law duties to decide what investments in the public good are also in the business interests of the company (Parliamentary Joint Committee on Corporations and Financial Services 2006; Corporations and Markets Advisory Committee 2006).

The Australian Government Corporations and Markets Advisory Committee (2006) advice on this issue is illustrative of the current ambiguous state of play. In affirming the obligation of the manager to make such decisions with business interests as the paramount concern, the Committee stated:

> *'The "business" approach does not involve inappropriate compromise or subordination by directors of the interests of shareholders to those of other interest groups. Rather, awareness of the relevant environmental or social considerations is part of any strategy to promote the continuing well-being of the company and to maximise value over the longer term … Consistent with the business approach, directors may sometimes choose to go further, where they see it as relevant to their business interests, in promoting particular societal values or goals or in seeking solutions to challenges facing their industry and the community. But this is not to suggest that companies bear some form of obligation to tackle wider problems facing society, regardless of the relevance of those problems to their own business.'*

As guidance for managers about how they ought respond to increasing expectations that the corporation will fund social good activities such as voluntary reporting, this advice is of limited value. The principles laid down by the courts maintain the dominance of a narrow view of a manager's duty as the custodian of shareholder's funds, but with the capacity to engage in social good activities such as voluntary reporting where a genuine business purpose can be found (Corporations and Markets Advisory Committee 2006).

The term triple bottom line (TBL) reporting was coined by Elkington in the 1990s (Elkington 2004).[5] It is widely used to connote non-mandatory reporting on economic, environmental and social activities, and arises from the development of corporate social responsibility (CSR) (Christen *et al.* 2006; Barut 2007; Environment Australia 2003; Raar 2002; Lamberton 2005). CSR generally refers to the idea that corporations have responsibilities beyond their legal duties that apply to a broad range of stakeholders (Corporations and Markets Advisory Committee 2006; Nolan 2006; Robins 2005; Suggett and Goodsir 2002). CSR has become closely associated with public reporting of the social and environmental impacts of corporations (Nolan 2006), to the point where terms such as 'corporate social responsibility', 'transparency' and 'triple bottom line' are almost interchangeable (Parliamentary Joint Committee on Corporations and Financial Services 2006). Related concepts such as social impact assessment (Vanclay 2004); social and environmental accounting and reporting (Moneva *et al.* 2006) and sustainability accounting (Lamberton 2005) are versions of social reporting.

Neither the business purpose test,[6] nor indeed the broader obligations in some other jurisdictions, provides the manager with specific principles for deciding whether, when or in what form to undertake non-mandatory reporting. The intent of this chapter is to provide specific guidance about the business judgement issues involved in balancing the interests of shareholders with the interest of the general public, with specific attention to the issue of voluntary reporting. The chapter will explore the dilemma, consider a specific real-world example of a corporate team facing this decision, and then outline a method that seems to offer a practical (or at least defensible) approach to resolution of the challenge.

What is in the business interests of the corporation?

In *The Social Responsibility of Business is to Increase Profits*, Nobel Prize winning economist Milton Friedman argues that the exclusive duty of business is to pursue profits (while refraining from engaging in deception and fraud). He states (Friedman 1970):

'... discussions of the "social responsibilities of business" are notable for their analytical looseness and lack of rigor. What does it mean to say that "business" has responsibilities? Only people can have responsibilities.'

His view is based partly on the belief that shareholders or other owners can themselves decide where they ought invest their own funds for social good purposes, allocating funds earned on their investment or from their own labour as they see fit. Friedman (1970) also states the executive:

'... would be spending someone else's money for a general social interest. Insofar as his actions in accord with his "social responsibility" reduce returns to stockholders, he is spending their money. Insofar as his actions raise the price to customers, he is spending the customers' money. Insofar as his actions lower the wages of some employees, he is spending their money.'

Friedman (1970) also argues that there may be an economic interest in devoting effort to community good, because this may generate goodwill and thereby profit. Friedman's argument is a more explicit statement of the duties of the manager than that made by the Commission cited above. Both suggest the same requirement to treat the business interests of the shareholders as the paramount concern. Implicit in both arguments is the view that managers who pursue non-business goals are likely to diffuse effort and focus, and that such social goal pursuit can be a mask for inefficiency, self-interest and non-accountability.

The alternative view, which seems to us to fly in the face of the moral component of fiduciary responsibility, has recently been put as follows:

'As a practical matter, as long as managers can plausibly claim that their actions are in the long-run interests of the firm, it is almost impossible for shareholders to challenge the actions of managers who act in the public interest.' (Reinhardt and Stavins 2010)[7]

The proper question, we would think, is not what can the manager 'get away with' in pursuit of their preferred goals with other people's money. Rather it is what ought the responsible manager do to properly engage with their legal and moral obligations.

The business case for voluntary reporting

Notwithstanding the conservative view (and amply demonstrated by other chapters in this book), expectations are growing that corporations will provide public reports about their impacts on society, environment and economy (the triple bottom line). These arise out of: 'good corporate citizenship'; public ownership reporting requirements; accounting standards, industry standards and international 'best practice' such as the Global Reporting Initiative.[8]

The business case for such reporting is made on various bases:

- the moral duty argument
- the transparency argument
- the customer trust argument
- the investor incentive argument
- the risk management argument
- the corporate culture argument
- the profit argument
- the social licence argument.

Business duties to society

Although there is a legitimate debate about the extent of this obligation, few would argue that executives do not have some duty to ensure the corporation acts in the interest of society (Christen *et al.* 2006; Barut 2007; Norman and MacDonald 2004; Adams and Zutshi 2004; Arnold and Day 1998). TBL reporting is a method to demonstrate that such responsibilities are being met. At least some corporations see good ethical standards as a benefit of TBL reporting (Parliamentary Joint Committee on Corporations and Financial Services 2006; Environment Australia 2003; Centre for Australian Ethical Research and Deni Greene Consulting Services 2003). A related argument is for transparency (Norman and MacDonald 2004; Centre for Australian Ethical Research and Deni Greene Consulting Services 2003; Deegan 2002) as a response to community demands for information on corporate social performance.

These arguments are given a commercial thrust by suggesting that there is a 'branding' benefit from consumer trust (Parliamentary Joint Committee on Corporations and Financial Services 2006; Suggett and Goodsir 2002; Adams and Zutshi 2004; Centre for Australian Ethical Research and Deni Greene Consulting Services 2003; Milne *et al.* 2005). This relates to the broader issues of social licence that are explored in other chapters in this book, and the narrower issues of consumer trust and brand loyalty. A further development is the increasing expectation that upstream suppliers in a product or market chain will 'sign on' to standards of social performance and reporting so as to enable the eventual retail marketer to claim laudable standards of social performance. Prime examples are the 'green' claims of major retail chains such as Coles and Woolworths in Australia, Tesco in the UK or Walmart in the USA (Bubna-Litic *et al.* 2001).

The rise of socially responsible investment (SRI) adds another driver for TBL reporting (Parliamentary Joint Committee on Corporations and Financial Services 2006; Christen *et al.* 2006; Nolan 2006; Suggett and Goodsir 2002; Adams and Zutshi 2004; Centre for Australian Ethical Research and Deni Greene Consulting Services 2003; Deegan 2002). Both investing and lending institutions are increasingly likely to require information on the social and environmental impacts of the entities into which they invest (Suggett and Goodsir 2002; Deegan 2002). Illustrative of this trend is that UK pension funds are now required to state their policies on socially responsible investment as part of their investment principles (Adams and Zutshi 2004).

Risk management is frequently mentioned in the literature as a reason to embrace TBL reporting (Parliamentary Joint Committee on Corporations and Financial Services 2006; Christen *et al.* 2006; Environment Australia 2003; Nolan 2006; Centre for Australian Ethical Research and Deni Greene Consulting Services 2003; Bubna-Litic *et al.* 2001). The risks generally cited are those mentioned above. It is also argued that public reporting is a bulwark against being 'caught out' by future legal and social obligations, sensitising the organisation to future requirements (Adams and Zutshi 2004). Increasingly these requirements do include mandatory reporting of environmental or social performance. For example under s 299(1)(f) of Australia's *Corporations Act 2001* (Cwlth), companies must now report compliance with environmental regulations. Engagement with non-mandatory accountability demands can assist organisations to evolve ahead of imposed requirements (Parliamentary Joint Committee on Corporations and Financial Services 2006) and may help them to head off the potential imposition of these (Parliamentary Joint Committee on Corporations and Financial Services 2006; Christen *et al.* 2006; Deegan 2002).

Many of these arguments are based on the corporate benefit of fostering a social values-driven culture, better communication and decision making that causes decision makers to expand the concerns they consider. It is argued that this leads to improvements in staff morale and employee commitment and thereby can assist in attracting and retaining staff (Parliamentary Joint Committee on Corporations and Financial Services 2006; Christen *et al.* 2006; Environment Australia 2003; Nolan 2006; Adams and Zutshi 2004; Centre for Australian Ethical Research and Deni Greene Consulting Services 2003; Milne *et al.* 2005; Bubna-Litic *et al.* 2001).

The business case is that corporations that deliver better social and ecological performance will be more profitable and that TBL reporting can help ensure that better social and environmental performance is achieved (Parliamentary Joint Committee on Corporations and Financial Services 2006; Environment Australia 2003; Norman and MacDonald 2004; Milne *et al.* 2005). TBL reporting arguably helps the company to demonstrate its trustworthiness, making it more likely that it can defend its reputation. Defence of the corporate reputation, and greater sensitivity to external expectations and impacts, also relates to the social licence to operate that is discussed in different forms throughout this book (Parliamentary Joint Committee on Corporations and Financial Services 2006; Christen *et al.* 2006; Deegan 2002).

Some of the argued benefits are contestable. Consumer decisions in practice do not always match consumer attitudes identified by research (Suggett and Goodsir 2002; Bubna-Litic *et al.* 2001). However, there is ample evidence of the capacity of consumers to exercise their economic muscle (Adams and Zutshi 2004). The investment incentive argument can also be contested because ethical investment represents a small percentage of total investments (Suggett and Goodsir 2002). The counter to this is that the extent of ethical investment varies between markets and, in almost all markets, is increasing (Parliamentary Joint Committee on Corporations and Financial Services 2006; Suggett and Goodsir 2002).

The dangers of boundary-less reporting

Even given concerns about misuse of corporate resources, the management of the risks of shifting community and legal expectations, the benefits of a healthy corporate culture and the protection of the social licence to operate are important considerations for any manager and a potential justification for a TBL reporting regime. However, for the responsible steward, the potential benefits from any investment must be weighed against the costs that will be incurred, and the risks that may arise from increased public reporting. If the costs and risks are substantial, then the benefits must be specific and substantial. However, if the dis-benefits are trivial, then the decision to provide extra information would not be contentious. For the next part of this chapter, we will consider the costs and risks that can come with public reporting.

What are the risks of voluntary TBL reporting? The obvious one is the risk of arming critics with the means to cause harm to the economic interests of the enterprise. The less obvious companion risk is that of misleading the public by hiding potentially damaging information. In trying to manage situations when the story that the data provides is less than flattering to the corporation, the manager is in a difficult position. One can easily envisage the manager faced with such a situation seeking the advice of the company's shareholder or public relations department to present the story in the most favourable light, hoping to ensure that the community sees only the positive side of the corporation's actions. The line between careful communications and deception might easily become blurred.

The term 'greenwash' is used to describe the practice of some corporations of masking their undesirable ecological impacts by skilful marketing and communications, or possibly by lying. Although many would say that this is risky because of the possibility of a backlash in attitudes to the corporation once the subterfuge is identified, far fewer would realise that there is a risk of significant penalties for doing so. Under the *Trade Practices Act 1974* (Cwlth) s 74, it is an offence to engage in misleading or deceptive conduct in trade and commerce. Recent cases and statements from the Trade Practices Commission make it clear that this provision does apply to misleading statements about environmental performance (Australian Competition and Consumer Commission 2008). Note that *misleading statements can include silence.* It is also a further offence to suggest that goods or services have sponsorship, approval, performance characteristics or other benefits that they do not have, and for breaches of this provision criminal penalties and fines of over $1 000 000 are possible. This limits substantially the freedom that the marketing department has to use its creative ability to present unpalatable facts in ways that suit the corporation.

The Trade Practices Commission requires that claims be both accurate and substantiable by objective evidence. It expects that claims will be specific, rather then general 'fluff', and be in plain English. They should not overstate the benefits, and even graphical tricks can be deceptive. The Commission points specifically to words like 'green', 'environmentally friendly' or 'environmentally safe', 'energy efficient' or 'recyclable' as potentially creating a legal risk.

These legal obligations are not without teeth. In one instance concerned with air-conditioners, the compensation package exceeded $3 000 000! Although the examples the Commission uses relate particularly to products, the legal principles apply equally to other forms of reporting about the attributes of the company. The decision to undertake voluntary triple bottom line reporting carries with it far more than a choice to add another task to the marketing department workload!

The risk of unconstrained cost

TBL reporting practitioners focus on determining what stakeholders want, and upon the forms in which this information might be delivered to best present the image that the corporation wants for itself. The desire for more information represents the 'demand' side in the market for TBL information. However, demand in normal markets is constrained by the price mechanism – if there is no price, then demand will consume a resource to exhaustion. No price boundaries are set on the demand for TBL reporting. Attempts to satisfy the demand can be expected to deplete resources beyond a reasonable level. The fable of the man, the boy and the donkey who seek to satisfy everyone's expectations, but end up losing all, is as relevant today as it was in Aesop's times.

The corporate sector has not developed clear principles for what ought be reported, when and why. The frameworks that have been used to advance the cause of TBL reporting are weak on principles to constrain the expenditures. They tend towards a focus on the 'how to report', rather than whether to report, treating reporting as an image issue and discounting the fiscal and risk aspects. Potential consequences of corporations not developing sound principles that place a constraint on the 'supply side' for TBL information include:

- The more information that is provided, the greater the demands from the public for more detail or different information.
- Corporations will carry a disproportionate financial and managerial cost of satisfying these demands. This represents a redirection of resources from meeting other clear accountabilities.
- The question of 'what accountabilities are we reporting against?' and 'what decisions do we intend to make?' will remain unresolved. This is a recipe for confusion, poor discipline and clumsy implementation.
- A lack of clearer principles will leave space for selective approaches to reporting, maintaining the appearance of accountability while 'fudging' the reality.

The consequences ought worry the responsible manager. These include:

- corporations investing money and effort in reporting things over which they have no effective control or upon which they have no real intention to act

- corporate expenditure on social good activities misdirecting shareholders funds and/or public resources into things that are irrelevant to the business purpose of the corporation
- corporations selecting their measures in an undisciplined manner. This may involve the moral hazard of choosing only to report the good news, treating TBL reporting as merely a public relations issue.
- the community being misled, and corporations failing to direct their energies into things that will make a substantial difference to TBL outcomes.

A case study on business-interest based TBL reporting

In the balance of this chapter, we will consider a case study about an anonymous corporation. This case study shows the extent of the reporting that is already required of a typical natural-resource-based corporation. It will illustrate the substantial effort and cost that is involved, and the potential complexities of adding another non-mandatory layer. It shows that responsible managers have sound governance reasons to be careful in deciding what additional information they might report, because the management complexities of gathering, analysing and reporting useful data are substantial. The case study also illustrates a business – purposes – based set of principles for deciding what to report. These reflect a strategic commitment to social reporting that marries the interests of the broader community, and the interests of the business.

The corporation is a water utility that operates within one state. It delivers water across a large area to rural and urban customers, and is government owned. It is a major corporation with a large number of employees and contractors, and billions of dollars in assets under management. We have designated this corporation as 'K Corporation'.

The case study was conducted over 3 months working with the corporation, which was considering its approach to GRI compliant voluntary reporting. The research team engagement was only for the period of development of the principles and approach to be used, not for implementation.

Interviews were conducted with K Corporation staff members who were responsible for aspects of its external reporting. These examined the external reports being generated, analysis required to produce each report, the sections of K Corporation that were performing these analyses and the source of the original data.[9] Part of the study involved developing a map of the current reporting. The reporting requirements placed on K Corporation are diverse. No single member of the organisation had comprehensive knowledge of the reporting taking place, nor had this information been collected in a document before.

Outputs from the interviews with K Corporation staff included maps of external reporting arrangements. This showed for each of K Corporation's external reports: the sources of information; the type of analysis performed; and where the report is sent. Coding used for the mapping exercise is illustrated in Figure 13.1, followed by the maps generated for various types of reporting (Figures 13.2 to 13.8).

A number of these reporting systems are required by law, with the requirements well specified. However, there is a surprising extent of non-mandatory reporting already being carried out by K Corporation. Both state and federal government departments require reports, and it is not always clear that these are mandatory, though such demands are treated as having legal status. There are also disclosures to local government; regional natural resource management authorities, customers and the general public, which are not legally required but are considered to be part of K Corporation's 'good citizenship'. These reports involve financial data, pricing information, water quality and use metrics, project updates and corporate plans. All these requirements are consistently fulfilled by K Corporation.

We were unable to quantify the costs of all of this reporting precisely, because the detailed costing and accounting work that this required was beyond the scope of the project. However,

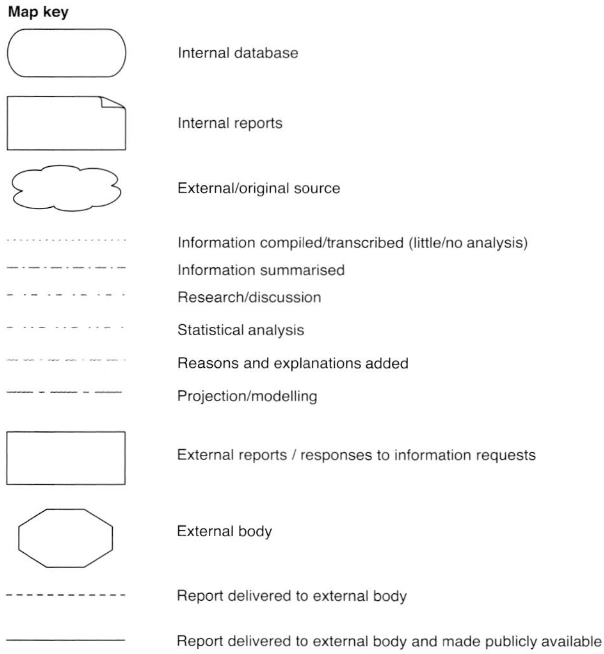

Types of report

Financial reports: Reports containing information such as profit/loss, and budget forecasts. Included in this type of reports are a number of financial statements to government departments.

Customer/pricing reports: Papers and presentations dealing with customer issues. Usually deal with pricing and allocation. Primarily produced for customers, but Government also informed. Includes discussion papers and brochures, as well as pricing lists.

Water quality reports: Indicators of water quality and information on health and safety issues delivered to Government Departments and/or Local Government. Includes reports on blue-green algae, the Safe Drinking Water Report, and nutrient/drainage reports.

Water use reports: Includes both formal reports, and requests for information from Government Departments and Local Government on water flows and usage.

Operational reports: Information on achievements and activities in regards to water. Contains reports on programs run for other authorities, responses to requests by Government for summaries of activities, Partnership Reports, and reports produced for bodies that have provided funding, detailing the use of the funding.

Corporate reports: Reports on activities as a corporation. Details current activities and future plans. Examples are: benchmarking reports, the Corporate Plan, and the Annual Report.

Note: Information flows from left to right. Vertical lines indicate information flows in both directions.

Figure 13.1: Coding used in maps

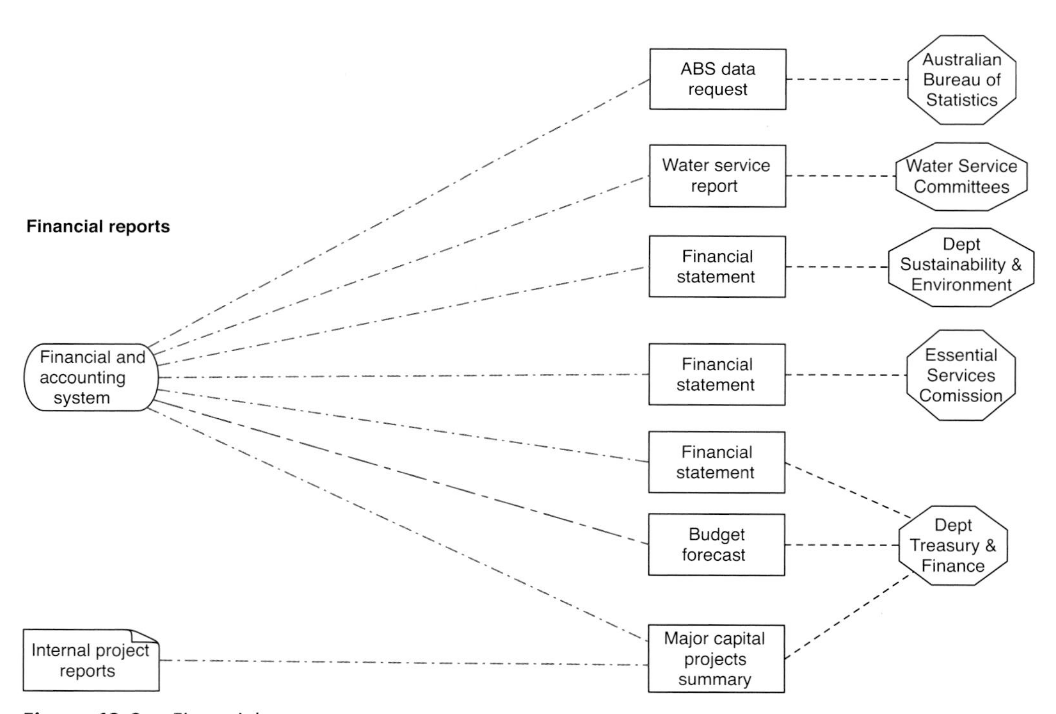

Figure 13.2: Financial reports

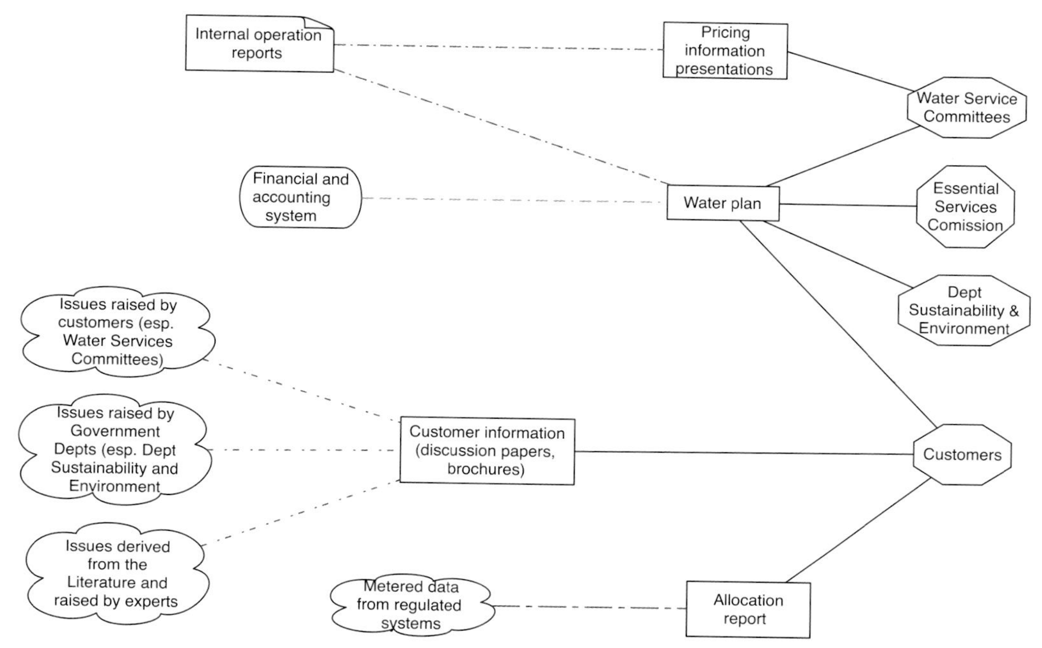

Figure 13.3: Customer/pricing reports

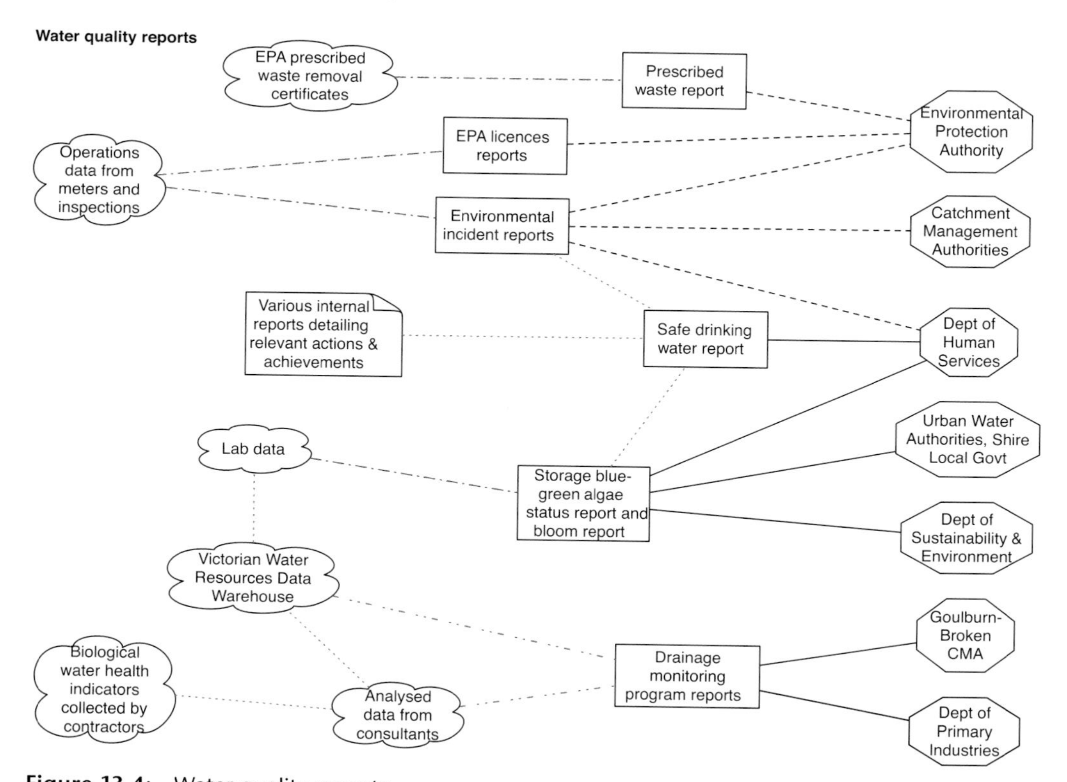

Figure 13.4: Water quality reports

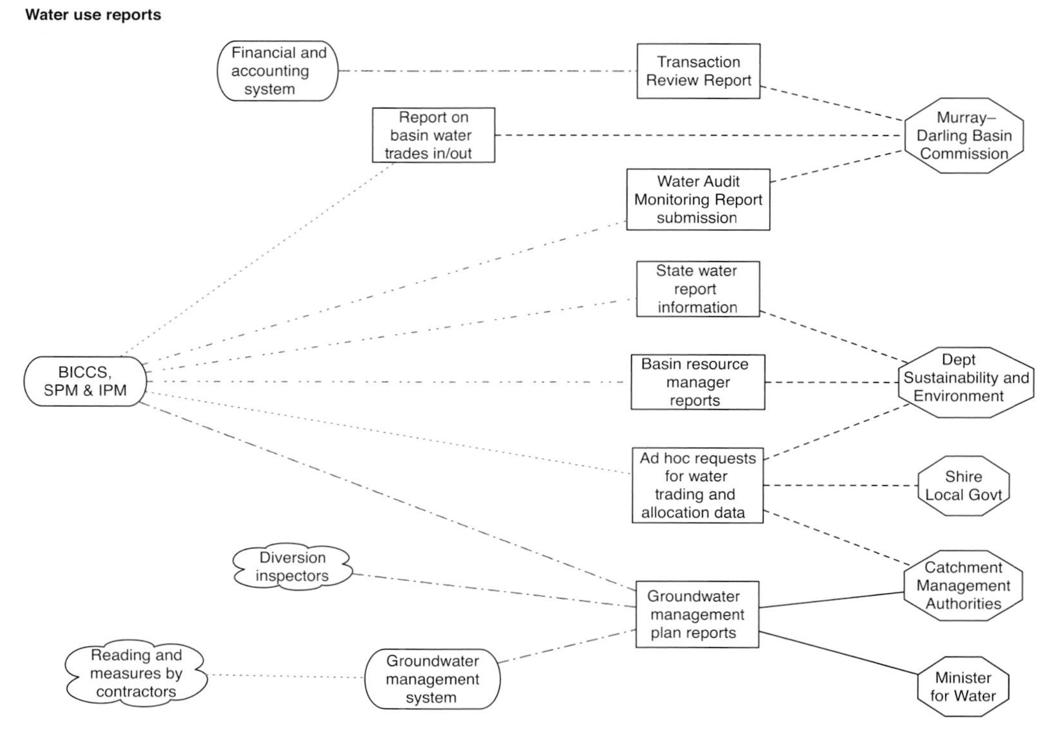

Figure 13.5: Water use reports

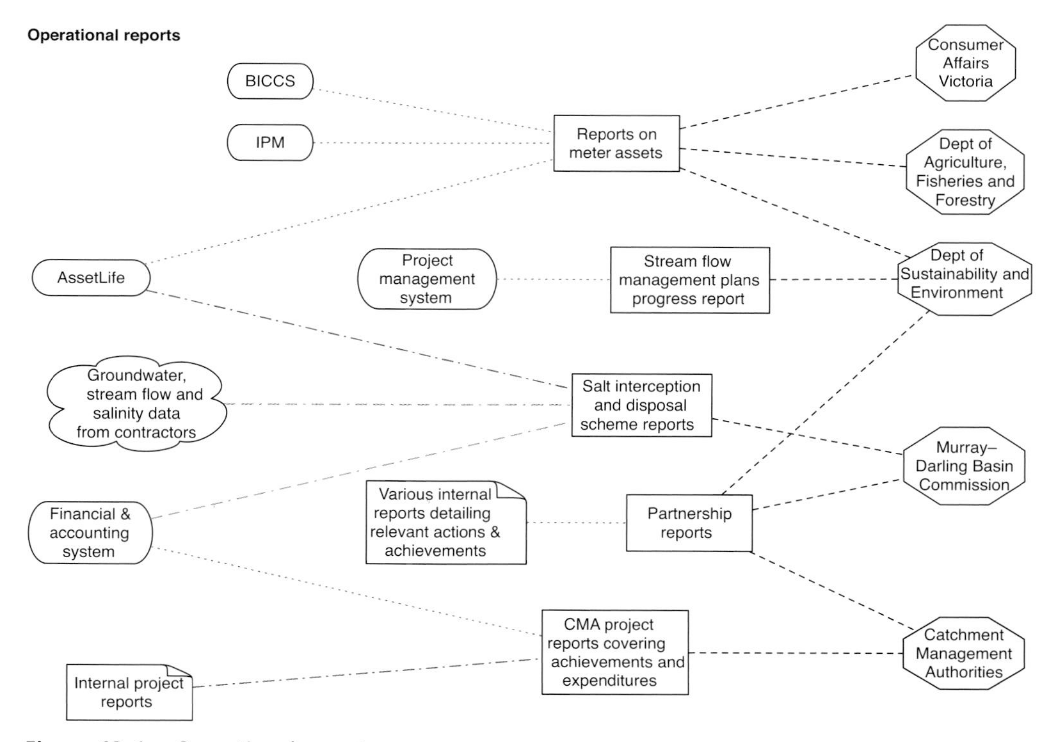

Figure 13.6: Operational reports

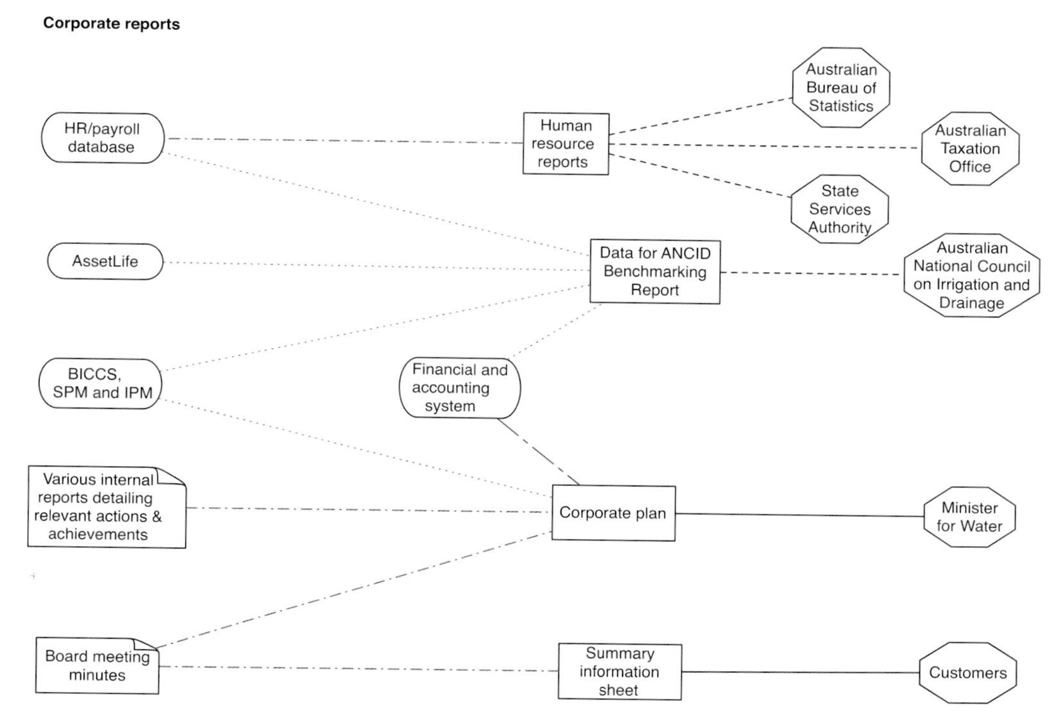

Figure 13.7: Corporate reports

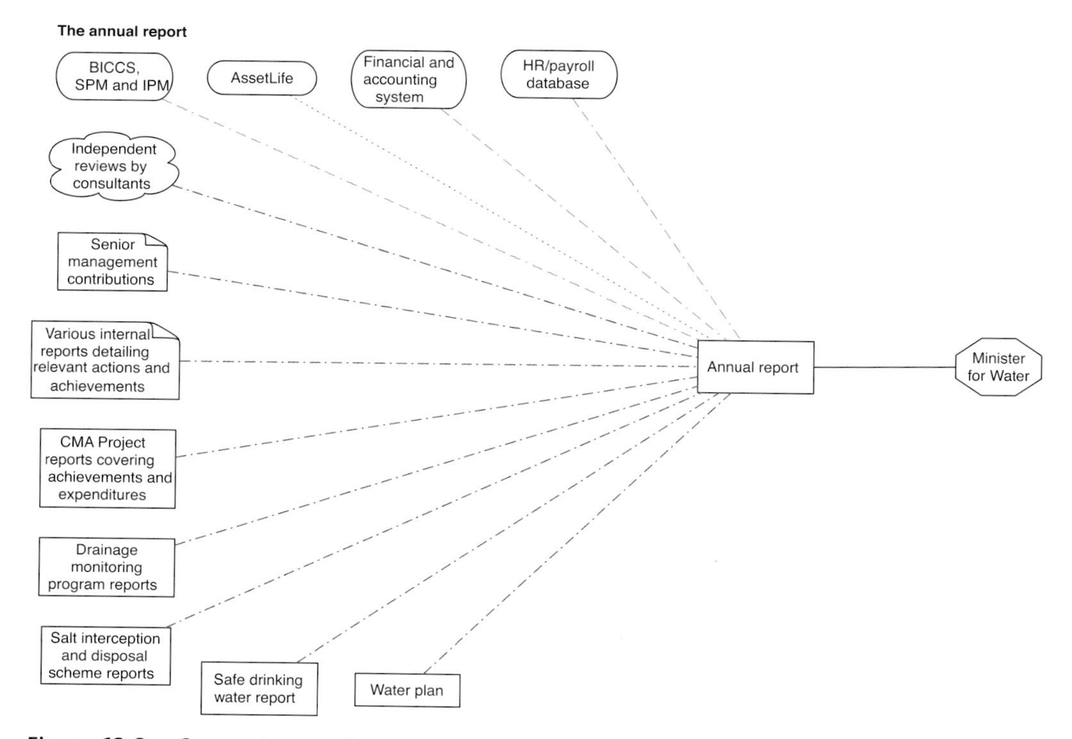

Figure 13.8: Corporate reporting

the costs of reporting (including reporting that is not strictly required by law) were estimated to be over $1.1 million in 2007.[10] K Corporation has no mechanisms to transfer the reporting costs to users. This means that there is no economic discipline over requests for information from other agencies, or from the community. The costs internalised by K Corporation are inevitably passed on to the users of water as additional costs, or to the broader community through reduced dividends to the government shareholder. There is a wealth transfer to those who want more information from the rest of the community who benefit from parsimony in the use of corporate resources.

The demand for further reporting to a range of (arguably legitimate) special interest groups is unconstrained by concerns for the cost to shareholders, even when the shareholder is the larger community. There is not a consensus in the community or the corporate sector about what or how to report, what information is relevant and (implicit in reporting) what action managers or owners ought to take on the basis of the information. Because such fundamentals of reporting are unclear, it is difficult to identify how to meet the requirements efficiently, or what boundaries there are on what to report, to whom, when and how. This lack of precision is in contrast to the clear accountability that exists for corporations under statutory, and even accountability under common law.

Towards a principle-based approach

It is possible to characterise reporting obligations at three levels, based on the degree to which the reporting expectation is formalised and specific. This taxonomy suggests some principles for determining which demands might be in the business's interest to satisfy.

Governance and shareholder accountabilities are legal obligations that managers have as custodians of the shareholder's funds, and are largely specified through statute. There are reports such as annual accounts and annual returns, taxation returns, various mandatory licensing information and a range of statutory lodgements, which must be met. These are the core reporting obligations where the management choice is about the most reliable/least cost method for satisfying the demand for information. Comparative audit and reporting performance standards are possible. The business purpose is compliance.

The second component (legal/managerial accountabilities, broadly characterised as corporate governance) is reasonably well defined in a legal sense, but there are limitations in the comparability and standardisation of reporting. Risk management accountability and reporting and related standards apply. Management has to determine what information is required to ensure good governance and, from this, may decide to share the information with the broader community or with other government or industry bodies. The management discretion is bounded by what the Board considers is necessary for governance purposes. The business purposes test is intrinsically part of the decision to make the investment.

The third component is information expectations from the public or other organisations where there is no mandatory requirement, or internal economic risk management or governance requirement. In this category, the satisfaction of the business purposes test ought require that there be a definable link between the provision of information and a specific and realistic business purpose; and the provision of this information ought satisfy a sound cost/benefit assessment.

There ought be principles underpinning a decision to report at the third level. It is proposed that a true TBL report should not merely be a restatement of matters that are covered in the two central fields of Figure 13.9 (regulatory requirements and governance of the resources of the corporation). It should deal with the external impacts of the corporation on its social, environmental and economic context, going beyond that which is otherwise reported. In the preceding

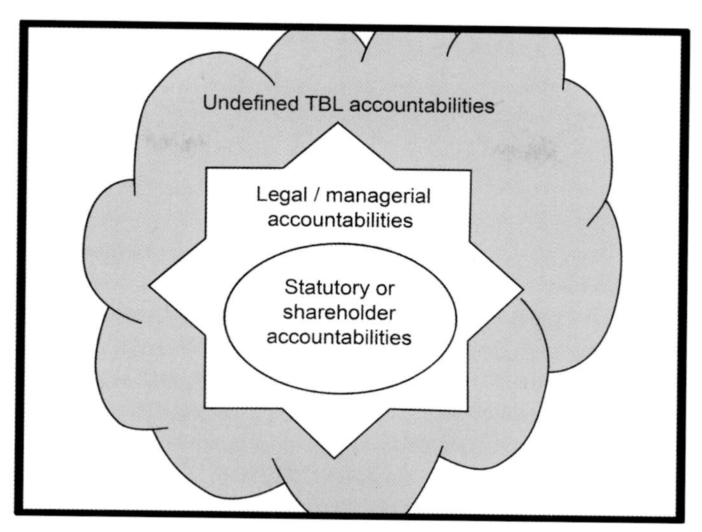

Figure 13.9: Levels of accountability

chapter, it has been shown that many TBL reports lump together information about investment, staff welfare and the like, but say little about the impact of the corporation outside of its core activities. Although this is a relatively easy type of report to prepare, it does little to satisfy the strategic objectives for voluntary TBL reporting, which we discussed early in this chapter.

An efficient approach to corporate reporting would, we suggest, report on matters where the organisation believes it has some accountability for outcomes and has some capacity to influence those outcomes. Merely reporting about the state of the communities within which the corporation works, or general information about the natural environment, does not satisfy this principle. An accountability discipline would limit the range of things measured and reported to those that are actionable, where the enterprise has determined it will take some action, or where the corporation believes its' activities might cause social or environmental harm for which it is likely to be held accountable legally or morally. It would distinguish between the content of legally required reports, internal or external governance reporting, and voluntary TBL reporting.

The practical implications of defining boundaries of reporting responsibility in this way are:

- focusing investment where accountability lies. This avoids fragmentation, waste of investment, and failure to deliver value to society.
- creating a basis for genuine allegiances and dialogue with relevant stakeholders
- sensitivity to society around issues where the organisation does have or may have some perceived responsibility
- ensuring that reporting, and particularly actions based on reporting, is established within the strategic management cycle, with expectations of action based on accountability.

The starting point is the question: 'what are the boundaries of the social, environmental and economic responsibilities of the corporation, outside of its boundaries of ownership?' This should define the matters where the organisation is prepared to adjust the economic interest of its shareholders (central accountabilities), staff, suppliers and customers (secondary accountabilities) to account for the interest of others outside this group (TBL accountabilities).

A principles-based approach to TBL reporting is likely to involve the corporation finding a balance between economic efficiency and community expectations. The gap that exists is the processes for the corporation to integrate community expectations and impact measurements with strategic business decisions about:

- the aspects of potential social accountability that are relevant to its strategy and the interests of its owners, workers and customers
- the aspects of social performance where it has the capacity and commitment to make a meaningful impact (which implies investing shareholder resources to at least some degree to make that impact)
- the business system changes that will be made to meet these commitments.

Implementation of the principles

After careful evaluation of the issues discussed above, and consultation with management in the case study corporation, the following implementation approach was developed. Its purpose was to close the gap between strategy and accountability for social, environmental and economic impacts outside the corporation. We believe that it is a sensible approach to applying the business judgement test to the decision to invest shareholder resources in voluntary reporting.

The implementation approach includes the following:

- Identify the priority areas where there is a corporate justification for prioritising investment in monitoring and taking action. This initial identification involves managers identifying the following three categories of impacts and issues (undertaken through a corporate survey):
  - economic, social or environmental impacts for which they expect legal or managerial obligations to emerge over the next 5 to10 years, and the likelihood of these requirements
  - economic, social or environmental effects that K Corporation has, or may have. Managers are invited to suggest where K Corporation has the potential to ameliorate harms or create additional social value, and possible strategies for doing so.
  - 'good citizenship' opportunities, where K Corporation has the potential to make a meaningful economic, social or environmental contribution without substantial deviation of resources from its core obligation to its owners, customers and staff.
- A team made up of senior managers, and potentially selected experts or stakeholders (implemented through a workshop), refine this list of issues as candidates for impact reporting given:
  - their potential emergence as legal or managerial accountabilities of K Corporation
  - their status as reflecting impacts that can be linked to the activities of K Corporation or their potential for K Corporation to demonstrate social leadership, either alone or in concert with others, without disproportionate diversion of resources from its main business.
- The resultant good citizenship and public accountability concepts are discussed with managers and selected stakeholders (nominated by K Corporation) to distil target priorities for:
  - GRI reporting
  - Social leadership action by K Corporation alone or in collaboration with other organisations.

    At this point in the process, surveys and engagement with staff and stakeholders can be used to deepen the corporation's understanding of expectations, and assist in refining the boundaries of its accountability.
- K Corporation will examine the least-cost, least-disruptive measures for reporting against potential accountabilities, and the internal processes required. This is an important

practical step, because the costs to the corporation arise largely from internal processes rather than external expectations.

- A management proposal is to be crafted for Board approval considering the following:
  - K Corporation's corporate citizenship approach and priorities for reporting and action. The boundaries of social, environmental and economic responsibility that are proposed are at least implicit in this proposal. This will consider expectations, capacity and corporate commitment to invest to have positive impacts. The aim is to provide clear guidance to management
  - the style and content of its public TBL reporting, in terms of issues and metrics to be used
  - the internal reporting approach to be used to satisfy this requirement.
- The senior management of K Corporation will draft a statement of its external accountabilities and leadership intentions, to form the cornerstone of its GRI reporting. This is intended to eventually become the foundation document for K Corporation's GRI reporting, which clearly stamps its commitment and approach to social responsibility;
- A Board-endorsed draft will be circulated for review by the owners and regulators of K Corporation, and selected stakeholder groups. This process will form the basis for the Board of K Corporation ratifying a formal corporate citizenship strategy to be reflected in its GRI reporting;
- A K Corporation statement, reflecting these elements, will be produced and made available publicly. It will be referenced in the GRI reports. This document should be subject to regular Board and management review, and remain a foundation for K Corporation's reporting.
- The K Corporation's annual report will thereafter be implemented with the GRI elements agreed by the Board and discussed with stakeholders, through reiterations of this process on an agreed cycle.

Conclusions

Corporate social responsibility and the protection of the social licence of the corporation to continue its commercial activities are important strategic matters. Some investment in these aspects of corporate strategy and practice is a justifiable use of shareholders' funds.

However, as will all other uses of corporate resources, good governance does require a disciplined and purposeful approach. This is often hard to see in the patterns of corporate social reporting, or even in public good investment and philanthropy programs.

We argue that a disciplined method, involving board decisions based on careful research and community dialogue, can bridge the gap between high aspirations and economic pragmatism. The method we outline is about creating a principled approach to balancing shareholder and community interest.

Corporate social responsibility expectations are unlikely to diminish. All the signs are that there are more likely to increase. Corporations that have a well-disciplined method for targeting their social responsibility investments and public reporting to their strategic needs ought be better positioned than their peers in responding to these challenges.

Endnotes

1 Section 181(1) of the *Companies Act* states that a director or other officer of a corporation must exercise their powers and discharge their duties:

(a) in good faith in the best interests of the corporation, and
(b) for a proper purpose.

2 Which requires that they act with the interests of the company (which transmutes into the interests of the shareholders as a whole) as their paramount consideration, and not to place other private interests over those of the corporation.
3 For a detailed discussion of Australian, UK and US approaches see Chapter 3 of Corporations and Markets Advisory Committee 2006.
4 For example, the UK Companies Act 2006, S172(1) a director of a company must act in the way he considers, in good faith, would be most likely to promote the success of the company for the benefit of its members as a whole, and in doing so have regard, (among other matters) to:
(a) the likely consequences of any decision in the long term,
(b) the interests of the company's employees,
(c) the need to foster the company's business relationships with suppliers, customers and others,
(d) the impact of the company's operations on the community and the environment,
(e) the desirability of the company maintaining a reputation for high standards of business conduct, and
(f) the need to act fairly as between members of the company.
5 This has been disputed by Norman and MacDonald 2004, 243–262.
6 For a specific discussion of the issues in the context of social responsibility see Reinhardt *et al.* 2008.
7 The interpretations of the legal obligations of management in that paper do not fit well with the statements of fiduciary responsibility made by Australian courts, though in practice the situation may not be far different than in the USA.
8 A case study of these reporting requirements for one water utility is provided later in this report.
9 Also sought were ideas from staff on the best means for increasing efficiency and approximate cost estimates for the reporting being done. This provided a start for considering the efficiency gains that could be made in current reporting activities.
10 The majority of this was labour costs of the corporation's employees.

References and further reading

Adams C and Zutshi (2004) Corporate social responsibility: why business should act responsibly and be accountable. *Australian Accounting Review* **14**(34), 31–39.

Arnold MB and Day RM (1998) *The Next Bottom Line: Making Sustainable Development Tangible*. World Resources Institute, Washington, DC.

Australian Competition and Consumer Commission (2008) *Green Marketing and the Trade Practices Act*. Australian Competition and Consumer Commission, Canberra.

Barut M (2007) Triple bottom line reporting: a study of diversity and application by Australian companies. PhD thesis. Swinburne University of Technology, Melbourne.

Bubna-Litic K, Leeuw LD and Williamson I (2001) Walking the thin green line: the Australian experience of corporate environmental reporting. *Environmental and Planning Law Journal* **18**(3), 339–350.

Centre for Australian Ethical Research and Deni Greene Consulting Services. (2003) 'The state of public environmental reporting in corporate Australia'. Environment Australia, Canberra.

Christen E, Shepheard M, Meyer WS, Jayawardane N and Fairweather H (2006) Triple bottom line reporting to promote sustainability of irrigation in Australia. *Irrigation Drainage Systems* **20**, 329–343.

Corporations and Markets Advisory Committee (2006) 'The social responsibility of corporations'. Corporations and Markets Advisory Committee, Sydney.

Deegan C (2002) Introduction: the legitimising effect of social and environmental disclosures – a theoretical foundation. *Accounting, Auditing & Accountability Journal* **15**(3), 282–311.

Elkington J (2004) Enter the triple bottom line. In: *The Triple Bottom Line: Does It All Add Up?* (Eds A Henriques and J Richardson) pp. 1–16. Earthscan Books, London.

Environment Australia (2003) 'Triple bottom line reporting in Australia: a guide to reporting against environmental indicators'. Department of the Environment and Heritage, Canberra.

Friedman M (1970) The social responsibility of business is to increase its profits. *New York Times Magazine* September 13.

Lamberton G (2005) Sustainability accounting: a brief history and conceptual framework. *Accounting Forum* **29**, 7–26.

Milne MJ, Tregidga H and Walton S (2005) *Playing with Magic Lanterns: The New Zealand Business Council for Sustainable Development and Corporate Triple Bottom Line Reporting.* Department of Accountancy and Business Law, University of Otago, Dunedin.

Moneva JM, Archel P and Correa C (2006) GRI and the camouflaging of corporate unsustainability. *Accounting Forum* **30**, 121–137.

Nolan J (2006) Corporate accountability and triple bottom line reporting: determining the material issues for disclosure. In: *Enhancing Corporate Accountability: Prospect and Challenges Conference*, 8–9 February, Melbourne.

Norman W and MacDonald C (2004) Getting to the bottom of triple bottom line. *Business Ethics Quarterly* **14**(2), 243–262.

Parliamentary Joint Committee on Corporations and Financial Services (2006) 'Corporate responsibility: managing risk and creating value'. Commonwealth of Australia, Canberra.

Raar J (2002) Environmental initiatives: towards triple-bottom line reporting. *Corporate Communications* **7**(3), 169–183.

Reinhardt FL and Stavins RN (2010) Corporate social responsibility, business strategy, and the environment. *Oxford Review of Economic Policy* **26**(2), 164–181.

Reinhardt FL, Stavins RN and Vietor RHK (2008) Corporate social responsibility through an economic lens. *Review of Environmental Economics and Policy* **2**(2), 219–239.

Robins F (2005) The future of corporate social responsibility. *Asian Business & Management* **4**, 95–115.

Suggett D and Goodsir B (2002) 'Triple bottom line measurement and reporting in Australia: making it tangible'. Allen Consulting Group, Melbourne.

Vanclay F (2004) The triple bottom line and impact assessment: how do TBL, EIA, SIA, SEA and EMS relate to each other? *Journal of Environmental Assessment Policy and Management* **6**(3), 265–288.

14

The duty of care: an ethical basis for sustainable natural resource management in farming?

Mark Shepheard

A *social licence* is about satisfying social expectations that go beyond the formal legal framework (Gunningham *et al.* 2002; Lynch-Wood and Williamson 2007). The concept of a social licence highlights that ownership of a legal right to resources does not guarantee community support for the exercise of that right. Rather the maintenance of a social licence depends on elements of law, beliefs, relationships, administration and expectations (Hone and Fraser 2004; Lyons and Davies 2007; Macintosh and Denniss 2004; National Farmers' Federation 2004; Raff 2005; Robertson 2003; Shine 2004; Spencer 2008; WWF 2005) Many aspects are inherently political, and not necessarily logical from the point of view of a farmer. The actual exploitative interest of the farmer can be a result of both well-defined property rights and poorly defined social expectations acted out in the form of restriction or expansion of the social licence to use that property. What constitutes a social licence can be difficult to specify because social expectations cover a diversity of concerns about economic, political, ecological, social and cultural consequences (Epstein 1987; Hutter 2006). The issues underlying social licence are often expressed vaguely and, for farmers, are couched as arguments about environmental stewardship and ecologically sustainable development (McKay 2006; Warhurst 2005). They do not provide precise practical guidance, are not constrained by legal rights or obligations, and do not necessarily respect private ownership (Lynch-Wood and Williamson 2007).

To illustrate, in irrigation farming, competing informal expectations can be seen in community debates about the volume and timing of water allocation, the impacts of irrigation on the environment, and the security of access to water by farmers and water for the environment. Such expectations reflect evolving social concerns about sustainability and the complex links that exist between water in its various uses, and community well-being (Cashman and Lewis 2007). These informal expectations go beyond formal accountability. Even more dramatic examples of the withdrawal of social licence are when communities punish actions that are legal but that violate community expectations (even if the community expectations are changing ones), such as with the recent conflict about mulesing (Australian Wool Innovation 2009; Lewis 2007; PETA 2007). In that instance, farmers found themselves facing a loss of local and international markets because consumers were convinced that particular sheep husbandry practices were cruel. Regardless of legality, and in order to regain the confidence of consumers, farmers have had to change their practices. Here the social licence to farm was powerfully demonstrated through the buyers' power to withhold access to consumer dollars. In this way, social licence can act to restrain rights of access and use.

In clarifying the region between what is stated as mandatory under statute and what is expected of farmers by the community, yet unstated, the concept of *duty of care* is anticipated to be most useful. It is relatively easy for everyone to agree that some boundaries should be in place, reflecting a distinction between behaviour that is 'accountable' and behaviour that is 'virtuous'. Moving from the abstract to the specific requires clear principles for doing so, and to date these are not in evident. Because expectations are not homogenous, and span the range from virtue to accountability, it is difficult to identify where these boundaries lie today.

Other chapters in this book have highlighted that the concept of a social licence invokes responsibilities to the community and the environment that go beyond readily specified property rights, and clearly specified legal obligations. One of the mechanisms that has evolved in an attempt to realign legal interest with social expectations of moral 'rightness' is to convert these expectations into a legal duty of care. This chapter explores the concepts that underpin this novel approach to bridging the gap between largely undefined social responsibilities and more specific legal obligations. Duty of care has been incorporated into a number of natural resource statutes in the belief that it will provide an effective tool for promoting farmers' sustainable use of natural resources, while at the same time providing greater certainty of legal obligations with less 'red tape' cost to primary producers. This broad ambition of the farming sector is linked to meeting social licence and formal legal requirements, but what is concealed are different expectations about what is meant by duty of care and what it might achieve. The expectations range from legally requiring virtuous behaviour by farmers (achievement of which may deserve to be rewarded, perhaps by improved access to resources) to expectations of minimal legal accountability (non-achievement of which may justify punishment, perhaps by denial of access). These different expectations will be explored in this chapter, beginning with a discussion of the nature of environmental responsibility in relation to farmers. We will consider how the changing nature of that responsibility is at the heart of the emergence of statutory duty of care in natural resource regulations. There is a fundamental question of whether a statutory duty of care is intended to legally enforce a minimum level of performance, or require virtuous behaviour that takes into account wider expectations about public responsibility. Both conceptualisations are argued in advocacy of a farmer's legal duty of care for the environment.

The statutory duty of care is promoted as a way to clearly define farmers' stewardship; that is, the boundary of responsibility between farmers and society about environmental protection of natural resources. A mixture of formal legal requirements and social licence expectations define these responsibilities. Formal requirements are documented in legislation and administrative instruments (such as regulations), which may reflect international agreements and treaties. Examples of formal instruments with expectations about farmers' natural resource management activities are: the *Native Vegetation Act 2003* (NSW), Property Vegetation Plans (NSW) and the RAMSAR convention.[1]

A statutory duty of care for environmental protection

Table 14.1 provides examples from Australian jurisdictions where a duty of care for environmental protection has been incorporated into legislation. These statutory versions are expressed in two ways: a brief form with the details imported by reference to a non-statutory code, or a detailed form fully expressed within the statute.

An example of the brief expression is the general environmental duty stated in the *Environmental Protection Act 1994* (Qld). This requires a person to take all reasonable and practical measures to prevent or minimise environmental harm. The Act also allows an industry code of practice to define the detailed meaning (s 436(3)). A code has been prepared by

Table 14.1: Australian legislation incorporating a ***duty of care*** for the environment

Legislation	Source of the duty
Environmental Protection Act 1994 (Qld)	s 319 General environmental duty
Land Act 1994 (Qld)	s 199 Duty of care condition
Catchment and Land Protection Act 1994 (Vic)	s 20 General duties of landowners
Natural Resources Management Act 2004 (SA)	s 9 General statutory duties and s 133 Specific duty to a watercourse
Environment Protection Act 1993 (SA)	s 25 General environmental duty
River Murray Act 2003 (SA)	s 23 General duty of care
Pastoral Land Management and Conservation Act 1989 (SA)	s 7 General duty of pastoral lessees
Environmental Management and Pollution Control Act 1994 (Tas)	s 23A General environmental duty
Forest Practices Act 1985 (Tas)	s 31(1)

the Queensland Farmers' Federation to detail how farmers can meet the duty of care. The Federation's code is approved by the Minister but does not represent a regulation made under the Act. The code centres on six 'expected environmental outcomes' (Queensland Farmers Federation 1998). These are that all reasonable and practical measures should be taken within the constraints of a sustainable agricultural system to:

- conserve representative samples of native species and ecosystems
- conserve the productive characteristics and qualities of the land and its soil
- conserve the integrity of waterways and the quality of water
- manage waste from on-farm activities
- conserve the quality of air through minimising the release of contaminants
- minimise the impact of noise on environmentally sensitive places at sensitive times.

Another illustration of the brief form is the statutory duty of care in Tasmania where the obligation is to prevent or minimise environmental harm or environmental nuisance (*Environmental Management and Pollution Control Act 1994* (Tas), s23A(1)). A code of practice is used to specify the requirements for compliance (See s 23(4) of the Act). *The Forest Practices Act 1985* (Tas) creates a code of practice for reasonable protection of the environment (See s 33(1) of the Act). The code describes a landowner's duty of care for the conservation of natural and cultural values, including measures to protect soil and water values and preserve other significant natural and cultural values (Forest Practices Board Tasmania 2000).

Although such guidelines exist, they do not determine what is 'reasonable': the detailed meaning of which may be disputed politically or in court; particularly these codes aim to supplant the implicit moral obligation connoted by the statutory duty with a series of technical procedures. For example, technical guidance about a landowner's duty of care under the Tasmanian Forest Practices Code is extensive for forest access, timber harvesting, conservation and management practices. It does, however, leave substantial discretion about how these practices will be implemented and remains silent on identifying 'to whom' and 'for what' the forest manager ought to be accountable (including specific legal obligations to the environment under other Tasmanian or Commonwealth legislation).

An alternate model for statutory expression of the duty of care exists in legislation from South Australia. The *Natural Resources Management Act 2004* (SA) contains a general duty to act reasonably in relation to the management of natural resources (s 9(1)). What this means is determined by reference to legislated factors to achieve: ecologically sustainable development (s 7(2)), reasonable measures (s 9(2)), and the statutory objects (s 7(1)). Detailed guidance in statues provides greater completeness in the legislation, but inevitably uses generic words and concepts, such as ecologically sustainable management, to define reasonable behaviour. Although the use of such terms is necessary, because they are intrinsic to care of the land, they can have multiple interpretations. This increases the practical complexity of definition for the duty holder.

The statutory duties of care identified above result from Parliamentary inquiry recommendations that have drawn on the common law duty of care, but have not critically questioned the function and meaning of such duty in the new context of environmental law (House of Representatives Standing Committee on Environment and Heritage 2001; Industry Commission 1998). Statutory versions of the duty of care focus on creating boundaries of responsibility that are adjudicated through an administrative process. Such a focus places a duty of care at the centre of a new ethic of natural resource stewardship (Gardner 1998). This is a virtue conceptualisation.

A statutory duty of care as a boundary of responsibility?

The unique role of the duty of care in the common law of negligence is to provide a mechanism to give legal effect to unstated expectations about how an individual ought to act, where their action might have an impact on others. It defines a limit to the freedom of the individual that is based on community norms. It does so by applying a careful process of logic to define what the citizen ought to have done, after the fact. From this citizens can infer their own code of behaviour for future actions.

This implies some level of social consensus that can be identified by a court about the boundaries of responsibility between the individual and broader community. However, a court in applying a legal duty of care determines such responsibility based on reasonable care (a minimum level of accountability). This is unlikely to meet the legitimacy and social trust concerns that are frequently required to defend the social licence (a virtuous level of stewardship performance). This signals a lack of consensus in the community that would be necessary to define what harm to the environment is actionable based on a duty of care. This can be contrasted with the civil and corporate uses of a duty of care, where it would seem that there is an understanding about the circumstances in which harm to another can lead to personal liability and about what actions might be considered 'reasonable'. The lack of consensus about the extent of farmers' stewardship obligations for natural resources and their responsibilities for environmental harm essentially reflects an unresolved debate about the responsibilities associated with access and use of property. By creating a legal duty of care, the Parliaments have potentially charged the bureaucracy and the courts with the responsibility for determining in particular instances what these social norms ought be. In so doing, they will be determining in part the effective extent of the rights to exploit which form part of a resource owners' property right.

Competing interpretations of property rights

Property is about the rules governing access to and control of resources. It exists as a relationship between people (the giver and receiver of the access right) (Stallworthy 2002) and between people and things that can be owned (which excludes other human beings). The relationship may be formally specified by rules governing access and use, including limits upon a property owner's freedom to exploit. The process of defining these bounds represents a framework through which ecologically and socially feasible behavioural norms are developed (Hajer

1995). However, there is a range of competing views about what the bounds should be, making it difficult to reach a consensus of norms about farmers' responsibility to the environment (Department of Sustainability and Environment Victoria 2008).

Benefiting from property requires that the community as a whole supports, and defends, an owner's 'right' to exploitation. For communities to invest in support and defence, they must feel comfortable that property owners will provide something beneficial to the community in return. That is, there must be a form of consensus about responsibility from and to the community (Martin and Verbeek 2002). Community expectations are not set, but are influenced by the law, and laws are largely set in response to community expectations. In a democracy, this relationship between the interests of the majority and the rights and freedoms of the individual lie at the very heart of the political system.

Where the community is dissatisfied with the bargain, it can either take away the 'right' or impose constraints through statutes, or it can apply force or sanctions to ensure that the collective interest is not ignored (Coyle and Morrow 2004; Raff 2005; Stallworthy 2002). It is normal for property rights to be subject to constraint. Land zoning, natural resource management legislation and industry or supply chain codes of practice are all partly expressions of the social consensus about responsibility. It is also a reality that the boundary between public and private interests is often implicit and it is not fixed across time. This means that there is always a degree of negotiation between property owners and the community about the true extent for the freedom to exploit private property.

The dominant paradigm in Australia has been that farmer responsibilities for management performance are a mixture of accountability created by specific laws and obligations to neighbours not to infringe their exploitative property right. This is a minimum form of accountability (Bovins 1998): an obligation to comply with the statutory and common law. Property is considered a largely un-attenuated right to exploit, with constraints imposed only where Parliament specifically makes clear this purpose or where exploitation may unjustifiably interfere with the interests of another property owner (or the physical well-being of other people). The emphasis is on minimum accountability for environmental protection with farmers' freedom to exploit their property rights being paramount (Fleming 1998). Within this traditional paradigm, the common law process for specifying reasonable care supports that freedom by specifying a level of accountability that is less onerous than the responsibilities that may be implicit in virtue-based expectations of the responsibility (Tucker LJ in *Latimer v AEC Ltd* [1953] AC 643).

Recent inclusions of duty of care into statutes dealing with natural resources, suggest a redefining of farmers' boundaries of responsibility (these are reviewed in greater detail below). A statutory duty of care seeks to import concepts of ethical responsibility or 'virtue' into farmers' legal responsibilities (Queensland Government 2007; Robson 1994; Victorian Government 2008). They implicitly define farmers as stewards of natural resources as part of the formal legal requirements.

It can be argued that citizens ought have a right to know what legal requirements do and do not allow (McBarnet and Whelan 1991). Unfortunately, the nature of a duty of care obligation, which must reflect changing social norms if it is to be effective, means that documented clarity of legal obligations is not feasible. However, society readily handles this lack of specificity in the common law use of duty of care, and so this would not seem to be an insurmountable problem.

Concealed tension surrounding farmers' stewardship

The rhetoric of the duty of care is interlinked with the language of 'stewardship'. Both those concerned with expanding the legal obligations of the farmer, and those concerned with restricting it, have adapted stewardship as a useful concept. The concept of stewardship is identified as the guardian of place, holding a position of responsibility (Pearsall and Trumble 2001). This has become important in modern conceptions of natural resource use and has been adopted in policies that advocate farmer responsibility for sustainable natural resource management (Barnes 2009; Carr 2002; Curry 2002). In relation to farming, the core duties of stewards are based on conservation to maintain resources for posterity and protection to save resources from harm (Barnes 2009). Such responsibility is said by critics to be lacking in modern agricultural production systems (Baldock *et al.* 1996). Instead, industrial agriculture has been blamed for causing environmental decay (Beale and Fray 1990; Cocklin 2005; Curry 2002; Fullerton 2001; Roberts 1995).

What is being advocated is environmental practices in which farmers do not deplete resources as part of their obligation to future generations (Royal Commission on Environmental Pollution 1996). Stewardship provides a conception of prudent or right behaviour to limit or reverse environmental harm (Lee 2005). Prudence is about ends, how to make important choices using a mixture of foresight, morals and self understanding, in effect a demonstration of virtue (Jacob 1995).

The discourse of stewardship seeks constraints on exploitation in the public interest. This paradigm anticipates the statutory duty of care as an effective way to define stewardship responsibilities. Stewardship acts to limit the exploitative freedom implicit in property rights. It suggests norms of conservation practice, and protection of legitimacy and social trust in return for environmentally and socially benign farming practice. It is a virtuous conception of performance supportive of social licence characterised by high conservation standards.

Figure 14.1 illustrates how this distinction between minimum accountability and virtue creates a tension between what political advocates expect a duty of care to mean and the meaning a legal boundary setting process will likely deliver. Such tensions have an impact on the expectations attached to farmers' natural resource access and use rights.

A failure to resolve the tension leads to a plethora of potential interpretations of the meaning of duty of care. The term hides competing expectations about defining legally enforceable boundaries of responsibility for farmers. The tensions will be brought into focus when its practical meaning needs to be defined (Cocklin *et al.* 2007; Crosthwaite 2001).

The alternative possible meanings of a duty of care

There are many different interpretations about what is a duty of care and what it can do to define stewardship obligations. These multiple meanings arise out of varied expectations for the term a duty of care. Twelve broad possibilities for what people mean when they talk of a duty of care can be distilled (Table 14.2), few of which reflect the use of a duty of care in negligence to define reasonable care. Many of the interpretations in Table 14.2 are used in debates, often without the conflicts between them being highlighted, creating a false sense of coherence between competing interests. In using the term a duty of care, advocates may be expecting quite different outcomes from its specific application.

These different concepts reflect opposing hopes of interest groups in their advocacy of the duty of care. These include:

- strengthening the property right and compensation claims of farmers
- strengthening the public interest claim to control farmers' management of natural resources
- creating new civil or government rights to intervene in the management of primary production

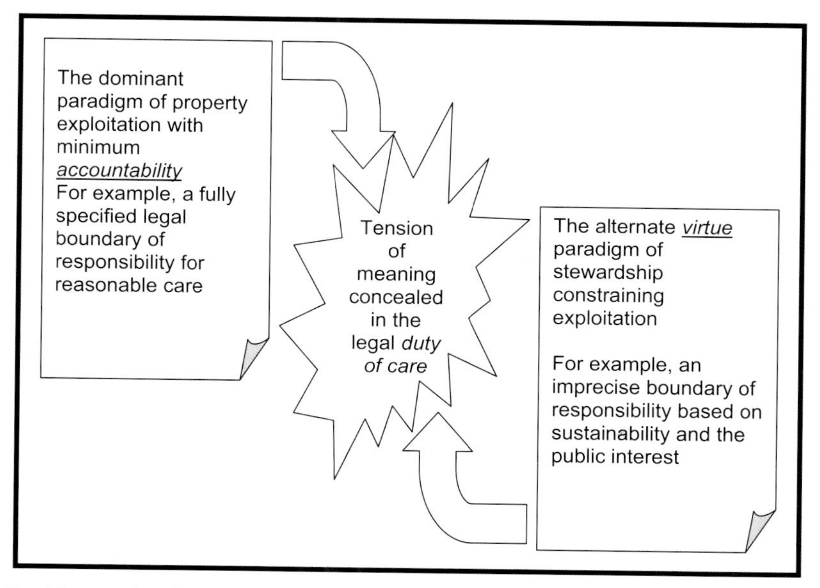

Figure 14.1: The tension between competing expectations for the duty of care

- strengthening 'right to farm' freedoms
- shifting of the costs of public good conservation from the private to the public purse
- embedding the costs of conservation as a cost of land tenure.

Clearly not all these competing expectations can be met. The potential for conflict and uncertainty remains high. Further refinement through the Parliament or judicial review will take time, and may impose high transaction costs (Baldwin and Cave 1999; Colby 1995; Dragun 1999; Guerin 2003; Martin *et al.* 2007; Maser and Heckathorn 1987; Palmer and Walls 1999; Richman and Macher 2006). Native vegetation regulation demonstrates that on-the-ground implementation of politically negotiated solutions to conservation conflicts can

Table 14.2: Possibilities for a duty of care and natural resource management by farmers

Potential interpretations of ***duty of care***	
Is it a flexible process for determining responsibility in a range of situations?	Or is it specific rules of practice that can be clearly stated?
Is it a method for handling disputes between individuals?	Or is it a method for determining compensation claims against the state for 'taking' of private resources?
Is its principal purpose to increase accountability for environmental and public good performance of private enterprise?	Or is it a means to safeguard resource use for private enterprise?
Does the term refer to a statutory duty of care, specified by Parliament?	Or does it mean a common law duty of care, developed by the judiciary?
Is it principally a tool used to frame political rhetoric?	Or is it a legally actionable concept with specific legal content
Is its purpose to define the collective duty of resource users generally across a generic range of circumstances?	Or is it intended to be a tool to evaluate individual performance in particular circumstances?

impose operating complexity and lead to loss for farmers, even for individuals who are not guilty of substantive harm-doing (Auditor-General of New South Wales 2006; Department of Infrastructure Planning and Natural Resources (NSW) 2004; Martin *et al.* 2007; National Farmers' Federation 2004; Productivity Commission 2004; Slee and Associates 1998; The Wentworth Group of Concerned Scientists 2003; WWF 2005).

An ongoing tension about stewardship and a duty of care

A statutory duty of care for environmental protection and stewardship of natural resources now exists in four Australian jurisdictions. These are worthy as general political principles of aspiration but do not, in their own terms, specify the practical meaning of the obligations they create. This is likely to lead to disputes where the problem of uncertainty of meaning and proof of breach will have to be resolved. These disputes may involve the courts and the application of common law principles.

The common law duty of care as a process for boundary formation views the facts of a situation against what would be considered reasonable behaviour in the same circumstances. Foreseeability of harm, the standard of care and consideration of common practice are relevant. Both in its civil use, and because of judicial standards of burden of proof in administrative law and statutory settings, a duty of care defines meaning based on minimum accountability, suggesting that minimal environmental protection is the most likely outcome of the statutory duty of care.

The effectiveness of stewardship is its effect on the exploitative freedom of property, forming norms of conservation behaviour, and providing legitimacy and social trust for environmentally benign farming industries. This suggests high conservation standards (a different meaning than 'reasonable care'). There appears to be a fundamental tension between what political advocates expect a duty of care to mean and the meaning a legal boundary setting process is likely to deliver. This important difference between legal accountability and social virtue is concealed by the words a duty of care.

The legislated duty of care throws some light upon the broader issues of social licence to operate are discussed in other chapters of this book. Social licence is a 'slippery' and political issue. Natural resource users would probably prefer to see their property rights as sufficient to guarantee their freedom to exploit what they have purchased, preferably in perpetuity. Other chapters in this book have suggested that in a world of increasing contests over scarce resources this hope will often be frustrated. Society can, and will, 'interfere', by redefining property interests, by creating laws that restrict what farmers can do, or by direct political or market place action. Only in some such instances will economic compensation be available.

The creation of a legal duty of care may provide a partial bridge between the uncertainties of moral and political claims and the property owners' expectation of freedom to exploit. However, that bridge can never be complete, precisely because of the ambiguities and inconsistencies of human values and interests. This chapter has demonstrated that the increased clarity of legal obligations that is sought by creating a legal standard of stewardship will necessary fall far short of resolving the challenges it is hoped to overcome.

Creating legal instruments of this sort is an important and positive move, if only because it creates an institutional setting within which the conflicts of values can be debated, and where some level of resolution will necessarily arise. However, this in itself will not secure the farmers' licence to operate.

Endnotes

1 The Ramsar Convention on Wetlands of International Importance especially as Waterfowl habitat, done at Ramsar, Iran on 2 February 1971. The English text of the convention is set out in Australian Treaty Series 1975 No. 48. The convention is adopted into Australian law by the *Environment Protection and Biodiversity Conservation Act* 1999 (Cwlth).

References and further reading

Auditor-General of New South Wales (2006) 'Performance audit. regulating the clearing of native vegetation. Follow-up of 2002 performance audit'. The Audit Office of New South Wales, Sydney.

Australian Wool Innovation (2009) 'Flystrike prevention in Australian sheep', <http://www.wool.com/Grow_Animal-health_Flystrike-prevention.htm>, accessed 4 December 2009.

Baldock D, Bishop K and Mitchell K (1996) 'Growing greener. Sustainable agriculture in the UK: report for the Council for the Protection of Rural England and the World Wide Fund for Nature'. CPRE and WWF, UK.

Baldwin R and Cave M (1999) *Understanding Regulation: Theory, Strategy and Practice.* Oxford University Press, Oxford, UK.

Barnes R (2009) *Property Rights and Natural Resources* (Studies in International Law). HART Publishing, Oxford, UK.

Beale B and Fray P (1990) *The Vanishing Continent. Australia's Degraded Environment.* Hodder & Stoughton, Sydney.

Bovins M (1998) *The Quest for Responsibility. Accountability and Citizenship in Complex Organisations.* Cambridge University Press, Cambridge.

Carr A (2002) *Grass Roots and Green Tape. Principles and Practices of Environmental Stewardship.* The Federation Press, Sydney.

Cashman A and Lewis L (2007) Topping up or watering down? Sustainable development in the privatised UK water industry. *Business Strategy and the Environment.* **16**(12), 93–105.

Cocklin C (2005) Natural capital and the sustainability of rural communities. In: *Sustainability and Change in Rural Australia.* (Eds C Cocklin and J Dibden). UNSW Press, Sydney.

Cocklin C, Mautner N and Dibden J (2007) Public policy, private landholders: perspectives on policy mechanisms for sustainable land management. *Journal of Environmental Management* **85**(4), 986–998.

Colby BG (1995) Regulation, imperfect markets and transaction costs: the elusive quest for efficiency in water allocation. In: *The Handbook of Environmental Economics.* (Ed. DW Bromley). Blackwell Press, Oxford, UK.

Coyle S and Morrow K (2004) *Philosophical Foundations of Environmental Law, Property, Rights and Nature.* HART Publishing, Oxford, UK.

Crosthwaite J (2001) 'Farmer land stewardship: a pillar to reinforce natural resource management?' *Connections* (updated 25/05/2007) <http://www.agrifood.info/connections/summer_2001/Crosthwaite.html>.

Curry D (2002) 'Farming and food. A sustainable future'. The Policy Commission on Farming and Food, London.

Department of Infrastructure Planning and Natural Resources NSW (2004) 'Draft native vegetation regulation 2004: regulatory impact statement'. Department of Infrastructure Planning and Natural Resources, Sydney.

Department of Sustainability and Environment Victoria (2008) 'Land and biodiversity at a time of climate change green paper'. Government of Victoria, Melbourne.

Dragun A (1999) 'Environmental institutional design: can property rights theory help?' University of Queensland, St Lucia.

Epstein EM (1987) The corporate social policy process: beyond business ethics, corporate social responsibility and, corporate social responsiveness. *California Management Review* **XXIX**(3), 99–114.

Fleming JG (1998) *The Law of Torts.* 9th edn. LBC Information Services, Sydney.

Forest Practices Board Tasmania (2000) 'Forest practices code'. Forest Practices Board, Hobart.

Fullerton T (2001) *Watershed. Deciding our Water Future. Juggling the Interests of Farmers, Politicians, Big Business, Ordinary People and Nature.* ABC Books, Sydney.

Gardner A (1998) The duty of care for sustainable land management. *The Australasian Journal of Natural Resources Law and Policy* **5**(1), 29–63.

Guerin K (2003) 'Encouraging quality regulations. Theories and tools: New Zealand Treasury Working Paper 03/24'. New Zealand Treasury, Wellington.

Gunningham N, Kagan RA and Thornton D (2002) 'Social licence and environmental protection'. Centre for Analysis of Risk and Regulation at the London School of Economics and Political Science, London.

Hajer MA (1995) *The Politics of Environmental Discourse. Ecological Modernisation and the Policy Process.* Oxford University Press, New York.

Hone P and Fraser I (2004) 'Extending the duty of care: resource management and liability'. School of Accounting Economics and Finance, School Working Papers Series, Deakin University, Melbourne.

House of Representatives Standing Committee on Environment and Heritage (2001) 'Public good conservation: our challenge for the 21st Century. Interim report of the inquiry into effects upon landholders and farmers of public good conservation measures imposed by Australian Governments'. The Parliament of the Commonwealth of Australia, Canberra.

Hutter BM (2006) 'The role of non-state actors in regulation'. Centre for analysis of risk and regulation discussion paper, London School of Economics and Political Science, London.

Industry Commission (1998) 'A full repairing lease. Inquiry into ecologically sustainable land management'. Industry Commission, Canberra.

Jacob BE (1995) Ancient rhetoric, modern legal thought, and politics: a review essay on the translation of Viehweg's 'Topics and Law'. *Northwestern University Law Review* **89**, Summer.

Lee M (2005) *EU Environmental Law: Challenges, Change and Decision-Making.* (Modern Studies in European Law, 6). Hart Publishing, Oxford, UK.

Lewis D (2007) 'Mulesing saves sheep from blowfly strike but is often seen as cruel'. *Sydney Morning Herald,* 22 March 2007.

Lynch-Wood G and Williamson D (2007) The social licence as a form of regulation for small and medium enterprises. *Journal of Law and Society* **34**(3), 321–341.

Lyons K and Davies K (2007) The need to consider the administration of property rights and restrictions before creating them. In: *Sustainable Resource Use: Institutional Dynamics and Economics.* (Eds Alex Smajgl and Silva Larson) pp. 208–219. Earthscan, London.

Macintosh A and Denniss R (2004) 'Property rights and the environment: should farmers have a right to compensation?' Discussion Paper Number 74, The Australia Institute, Canberra.

Martin P and Verbeek M (2002) Property rights and property responsibility. In: *Property: Rights and Responsibilities.* pp. 1–12. Land and Water Australia, Canberra.

Martin P, Bartel RL, Sinden JA, Gunningham N and Hannam I (2007) 'Developing a good regulatory practice model for environmental regulations impacting on farmers'. Land and Water Australia, Canberra.

Maser SM and Heckathorn DD (1987) Bargaining and the sources of transaction costs: the case of government regulation. *Journal of Law, Economics and Organisation.* **3**(1), 69–98.

McBarnet D and Whelan (1991) The elusive spirit of the law: formalism and the struggle for legal control. *The Modern Law Review* **54**(6), 848–873.

McKay J (2006) Issues for CEO's of water utilities with implementation of Australia's water laws'. *Journal of Contemporary Water Research and Education* **135**, 115–130.

National Farmers' Federation (2004) 'Policy on sustainable production, land and native vegetation'. National Farmers' Federation, Canberra.

Palmer K and Walls M (1999) 'Extended product responsibility: an economic assessment of alternative policies'. Resources for the Future, Washington.

Pearsall J and Trumble B (Eds) (2001) *Oxford English Reference Dictionary.* 2nd rev. edn. Oxford University Press, Oxford.

PETA (2007) 'Australian wool farmers "strike" again', *The PETA Files Australian Wool Archive* <http://blog.peta.org/archives/australian_wool>, accessed 4 December 2009.

Productivity Commission (2004) 'Impacts of native vegetation and biodiversity regulations'. Productivity Commission, Melbourne.

Queensland Farmers Federation (1998) 'The environmental code of practice for agriculture'. Queensland Farmers Federation, Brisbane.

Queensland Government (2007) 'Delbessie Agreement: state rural leasehold land strategy (December 2007)'. Department of Natural Resources and Water, Brisbane.

Raff M (2005) Toward an ecologically sustainable property concept. In: *Modern Studies in Property Law.* (Ed. Elizabeth Cooke) pp. 65-90. HART Publishing, Oxford, UK.

Richman BD and Macher JT (2006) 'Transaction cost economics: an assessment of empirical research in the social sciences'. Duke Law School Legal Studies Paper 115, Duke University, Durham, USA.

Roberts B (1995) *The Quest for Sustainable Agriculture and Land Use.* UNSW Press, Sydney.

Robertson S (2003) 'Hon. S. Robertson, Property rights, responsibility and reason'. Legislative Assembly Queensland Parliament, Brisbane.

Robson MJ (1994) Queensland *Parliamentary Debates* in Legislative Assembly Queensland Parliament (Ed.), *Environment Protection Bill Second Reading Speech* (Hansard).

Royal Commission on Environmental Pollution (1996) 'Sustainable use of soil'. HMSO, London.

Shine C (2004) 'Using tax incentives to conserve and enhance biological and landscape diversity in Europe'. STRA-REP, United Nations Environment Programme, Strasbourg.

Slee, Denys and Associates (1998) 'Remnant native vegetation, perceptions and policies: a review of legislation and incentive programs'. Environment Australia, Canberra.

Spencer P (2008) 'Australian sustainable farming for the 21st century: ecosystem services protocol', <http://www.sosnews.org/peterspencer/>, accessed 5 July 2008.

Stallworthy M (2002) *Sustainability, Land Use and Environment. A Legal Analysis.* Cavendish Publishing Limited, London.

The Wentworth Group of Concerned Scientists (2003) 'A new model for landscape conservation in New South Wales'. WWF, Sydney.

Victorian Government (2008) 'Green paper: land and biodiversity at a time of climate change 2008'. Victorian Government, Melbourne.

Warhurst A (2005) Future roles of business in society. The expanding boundaries of corporate responsibility and a compelling case for partnership. *Futures* **37**, 151–168.

WWF (2005) 'Native vegetation regulation: financial impact and policy issues'. WWF Briefing Paper, WWF, Sydney.

15

Co-management as a social licence initiative

Claudia Baldwin, Mark Hamstead and Vikki Uhlmann

This chapter presents the potential for managing water using co-management as a means for irrigators to be involved with developing, monitoring and enforcing rules for water sharing. The discussion explores examples of co-management, which indicate the potential of self-governing entities in monitoring and compliance and the potential role for co-management in Australia. Such an approach has the potential to allow irrigators to demonstrate good stewardship, in part supporting their credentials as justifying their *social licence* to use and manage water.

What is co-management?

One method of managing water as part of a social licence approach is through 'co-management', whereby irrigators are involved with developing, monitoring and enforcing rules for water sharing. The literature also refers to this concept as 'self-management', 'self-regulation' or 'self-governance'. However, from our perspective, the term 'co-management' more suitably captures the intention of shared decision making (preferably equitably) between government and resource users for collaboratively managing sustainable resource use (McCay and Jentoft 1996). An important feature is that co-management connotes a form of partnership between the state and the resource users whose actions are intended to be regulated and supervised.

Substantial literature suggests that resource users can effectively manage common pool resources, such as water, within a nested framework supported by government (Ostrom 1990; McKean 1992; Bromley and Cernea 1989). Based on hundreds of local and regional case studies from around the world, principles and models for self-governance of common-pool resources, such as groundwater, have been developed (Ostrom *et al.* 1999; Ostrom 2005; Ostrom and Nagendra 2007). Ostrom and colleagues identified eight basic principles that can support cooperative behaviour. The first three are:

- *Clearly defined boundaries.* The people holding access rights are clearly defined, as is the territory to which the rights apply, providing certainty about what is being managed and for whom.
- *A match between contribution and benefits.* Community-based management requires investment of time, effort and often, financial resources. To be long-enduring, participants must feel that they receive an equitable benefit from their contribution. Processes must be designed to ensure fairness.
- *The right to make and adjust rules.* In many of the successful regimes studied, the right to make changes to the rules under which the system operates was limited to the members: those people affected by the changes. Users who feel that rules or changes are not made in

response to local conditions, or do not match their needs, may simply ignore them. Any changes to the system must make sense in order for members to continue investing in it.

These three principles deal with institutional operational rules. The following three suggest a substantial departure from a typical top-down style of resource management in terms of monitoring, sanctions and conflict resolution:

- *Monitoring.* Those that monitor are either the members/users, or they report back to the users. Users decide what information is useful and to be collected. When users are genuinely engaged in decisions about rules affecting their use, there is greater likelihood of them following the rules and being prepared to monitor others, than when an external authority simply imposes the rules.
- *Graduated sanctions.* Infringement of rules occurs in any system. Successful co-management systems were found to have a graduated system of sanctions, from a warning for a first contravention, through to loss of access to water or loss of membership for serious or continual disregard of the rules.
- *Conflict resolution.* As with infringements, conflicts are inevitable, particularly when there may be ambiguity in rule interpretation or a rapidly changing system. Successful systems used local methods with limited costs to resolve internal conflicts, frequently as part of a leader's role. However, conflict resolution mechanisms can be informal or more formal with decisions being recorded.

The last two principles are features external to the body – recognition and nesting.

- *Recognition of the legitimacy of the self-management system by government authorities and by other resource users.* Without formal recognition, it is difficult to sustain cooperative behaviour and commitment by the range of parties required for appropriate resource management.
- *Nested enterprises.* Some, but not all, local bodies may be nested within regimes that operate at larger scales such as regional or state, levels which can provide additional support (Weinstein 2000).

Drivers for co-management in Australia

These principles can inform the design of community-based management of water resources. This chapter outlines seven case studies, which feature a diverse range of characteristics of community-based governance (Baldwin *et al.* 2008) and an attempt to apply Ostrom's principles to a proposal by irrigators for co-management in the Lockyer valley in Queensland (Baldwin 2008).

These studies found the drivers for irrigators proposing community-based management of water resources in Australia were:

- seeking to avoid government 'interference'
- values of self-sufficiency, self-determination, and managing one's own destiny
- concerns about additional costs due to water pricing and costs of metering, particularly with introduction of water reforms
- the need for flexible management of enterprises to respond to business and crop needs
- the need for good resource management information, which might be derived through collaborative arrangements
- seeking efficiencies through collaboration with those with common interests and needs
- a collaborative search for alternative or additional water supplies
- seeking equitable application of rules because participants develop and implement them.

Case studies of community-based governance arrangements

This research involved seven case studies of community based governance arrangements in five Australian state jurisdictions. These were: First Mildura Irrigation Trust (FMIT) in Victoria; Central Irrigation Trust in South Australia; North and South Burdekin Water Boards in Queensland; Abercrombie Pumping Association, Bega Valley Water Users Association and West Corurgan Private Irrigation Board (all in New South Wales); and the Winnaleah Irrigation Scheme in Tasmania.

The analysis involved a desk-top review of legislation, policies, annual reports and other documents relevant to each case study. This information was supplemented by telephone interviews with government agencies and members of the community-based water management organisations. The researchers used a consistent set of questions and approach focused on the nature and operation of the organisations and perspectives on successful aspects and areas for improvement (Baldwin *et al.* 2008).

The case studies demonstrated large and small memberships and three entity types: an entity defined in water legislation; a company limited by guarantee; and a non-profit incorporated association. All were governed by a Board or committee of six to ten elected from the membership. One had representatives from local government and a local industry. All were well defined (principle 1) and considered 'legitimate' by government (principle 8). The type of entity and structure was not correlated with the roles and responsibilities or delegated powers of entities. For example, one non-profit organisation had delegated responsibilities equivalent to statutory water entities, because that was permitted under its state legislation. All of the entities operated in a 'nested framework' to some extent.

Most of the entities exist for the purpose of supplying surface water to members for irrigation. A number of the entities distribute water as a bulk entitlement holder to licensees according to defined entitlements and manage water trading according to the water-sharing plan developed by government (with some involvement of stakeholders). Only the Burdekin Boards have a distinct resource management purpose, which is to replenish the aquifer.

In relation to Ostrom's fifth principle, most of the case study entities had the ability to impose penalties and/or cut off supply if members did not comply with allocation rules. In large organisations, where staff is employed to monitor and ensure compliance, this is appropriate. However, in smaller groups, a preferred approach for self management entities might be for them to have the ability to warn the member, then report illegal take to a government agency for enforcement, thus separating the compliance role.

All of the case study entities were supported by fees or levies on irrigators, which were based on a flat membership fee, unit of entitlement, area used for irrigation or volume of water used. The level of charges was related to the responsibilities of the entity and the number of staff needed to discharge the responsibilities. The largest had a budget of $8.2 million and a staff of three, while the smallest had a budget of $3000 to cover administrative costs. Their input was directly related to their responsibilities (principle 2).

Success factors of the case study entities were identified as being that:

- they have a clear purpose, which the entity members understand and recognise as being mutually beneficial
- they have stable, participatory, governance arrangements
- the members are willing to pay for the services the entity provides.

Although some of the entities were able to promote water use efficiency through training and grants for improved infrastructure, there were concerns that some entities protected their members' current extraction levels rather than seeking sustainable resource management.

From government's perspective, this is a serious issue because government is accountable to protect environmental assets and provide other public benefits, under the National Water Initiative (NWI).

It is clear from the research that community-based governance entities can form and fund themselves. They do not need the permission of government. However, government can legitimise and support their roles by recognising them in legislation, delegating powers, engaging them in development of plans, sharing data and by providing support for training and transition activities. Such entities can also foster credibility and leverage resources by collaborating with government, research agencies and regional natural resource management bodies. In relation to Ostrom's principle 3, though, cases where government has completely delegated water sharing, rule making and administration to a group of self-supply water users appear to be non-existent in Australia. The detailed case study that follows highlights some of the complexities involved in seeking this level of delegated co-management.

Proposal for co-management in the Lockyer Valley, Queensland

During 2005 and 2006, Lockyer Valley irrigators (LWUF) developed a proposal for 'co-management' of the entire valley through a series of facilitated meetings. Some involved the water management agency, Department of Natural Resources and Water (NRW) (Baldwin 2008). The initial LWUF proposal was that government would provide an overall allocation to each of the 18 management areas within the Lockyer, formalised in the Resource Operation Plan (ROP). All irrigation bores would be licensed, metered and monitored. A Board of Management comprised of LWUF members, South East Queensland Catchments (SEQC) and possibly government would administer the system, with meters owned and monitored by the Board (LWUF 2006).

Irrigators in each management area would work together using information on their aquifer supported by computer modelling and bore monitoring data to determine appropriate arrangements for use. Aquifers were not permitted to be extracted beyond a certain level. Seasonal flexibility at property level would be negotiated within each management area. It could include provision for a temporary, but not a permanent, trading system, retaining long-term control of the resource within the Valley. LWUF agreed to independent oversight and compliance management by government. In summary, the LWUF proposal was for a system of small management zones, nested within an overarching governance framework under the auspices of a LWUF Board, which would be responsible for managing an allocation or arrangements as specified under the Water Resource Plan and Resource Operation Plan and possibly through contractual arrangements between all parties. The proposal was designed to be consistent with Ostrom's rules for self-governance.

In spite of the series of negotiations with NRW, the initial proposal was not accepted. Government decided it could not sit on the proposed Board because of its potential conflicts with compliance role (LWUF 2006). It would participate in a Technical Advisory Committee and thus referred to the proposal as a 'regulatory self-management model' (NRM 2005). Although the term 'co-management' usually describes shared decision making between government and resource users (McCay and Jentoft 1996), not having government on the Board could still be consistent with the concept of nested governance (Sarker *et al.* 2009).

Secondly, NRW did not agree to allocate water by volume to self-management units. It decided it would need to provide individual licences. Although NRW has declared the entire Valley as a sub-artesian area, it did not agree to apply co-management across the Valley. Government focused on the Upper Lockyer because other parts of the system were already regulated, thus placing landholders under different management regimes. From an LWUF perspective, this did not recognise the connection between Upper, Central and Lower catchment aquifers. It also continues the inequitable approach, which had allowed unfettered access to groundwater for those in the Upper and Lower Lockyer, yet restrictions on take in the Central. A system that is not seen as equitable and fair is unlikely to be enduring (design principle 1) (Sarker *et al.* 2009).

A fourth issue between LWUF and government was the ownership of meters. LWUF wanted to own, maintain and monitor meters on farm bores. The government's state-wide metering policy requires that meters are owned and monitored by NRW. This was important to irrigators who wished to monitor directly and more frequently than NRW to better understand the relationship between water use and aquifer levels, to be independent, and to show responsibility for land and water use. They also thought it would be less costly and that meter installation could occur more quickly. In spite of areas of disagreement, both parties agree that there should be meters and that compliance should be independent (LWUF 2006).

Ostrom's work suggests that effective water resource management is more likely to occur if rules are based on sound data and developed in collaboration with users. Involving entities and members in monitoring extraction and resource condition provides continuous feedback about the resource. There will always be levels of uncertainty. Therefore it is essential that users understand the system characteristics, impact of extraction and these uncertainties so that an adaptive management approach is adopted.

Although the Lockyer lacks the history of formal collaboration on groundwater irrigation that one finds in many surface water irrigation systems, 'neighbourliness' is a practised norm (Baldwin 2008). The need to cooperate, forced by water scarcity and backed by an overall sense of community, could have provided a basis for collaboration among irrigators. It could have recognised their expertise and knowledge and allowed flexible management within overall constraints of the water-sharing plan. If effective, it could have reduced the operational demands on government. Co-management would have met the expressed and core needs of irrigators to have some measure of control over their own future (Baldwin 2008, Fisher *et al.* 1991). Influencing the rules and overseeing operational arrangements is empowering for irrigators. Thus, an opportunity to introduce co-management and commit to collaboratively manage the resource with government was lost. Why this was so suggests some challenges, which need to be overcome for effective co-management.

Barriers and risks of co-management

A number of barriers prevent co-management from being adopted, based largely on perceived risks. Nevertheless, measures can be put in place to manage these risks, as illustrated in Table 15.1.

To cater for the risks from both parties' perspectives, community-based governance needs to be formalised, rules (a plan) established in a transparent and inclusive way and responsibilities accepted and funded within the capacity of the irrigators to support them. Leadership needs to encourage collaboration among irrigators, and between irrigators and government.

Co-management and *social licence*

A social licence granted by the community is based on community perceptions of legitimacy through conformance with formal and informal legal, social, and cultural norms (Thomson and Joyce 2008). This can require time to be allocated to engage all voices. Just because an entity achieves legitimacy, for example, through obeying all laws, it does not necessarily mean that it will be socially accepted. A perception that entities will meet expectations and do what they say they will do is essential to credibility. Third-party review and evaluation of the language adopted and implementation and compliance will assist in overcoming the

Table 15.1: Barriers, risks and possible measures

Barrier	Perceived risk	Contingency measure
Government resistance to devolving responsibility	Entity not considered legitimate because of perceived lack of irrigator support and lack of capability.	Entity to employ skilled person to coordinate irrigators and manage operations
Government resistance to devolving responsibility	Entity not meeting obligations and therefore government not meeting NWI commitments. Perception that government can do better.	Supportive framework that includes agreements between entity and government, which identify roles and responsibilities for monitoring and compliance. Develop water plans collaboratively to ensure common understanding of environmental requirements and irrigator needs.
Credibility of organisation: acceptance by irrigators of the entity	Lack of trust and collaboration.	Professional advice on governance and organisational operating rules. Employ skilled coordinator.
Size of entity (membership)	Financially unviable – members unable to support part-time staff or entity beaks up.	Roles and responsibilities should be tailored to the capacity of the entity. Contingency agreement between entity and government.
Cost effectiveness of co-management	Benefits not worth the expense.	Reconsider benefits; e.g. cost-effective purchase of water efficiency equipment; eligibility for grants or other assistance. Entity to collaborate with government in developing rules, with individual responsibility for implementation.
Complexity of business that includes metering, trading and pricing	Additional skills and knowledge needed.	Government to take responsibility to ensure collaborative water planning processes that are transparent and inclusive, to build understanding and commitment. Develop irrigator capacity. Employ appropriately skilled people.
Difficulties with enforcement in self- supply	Non-enforcement of non-compliance. State not meeting NWI commitments; environmental benefits not met. Community unrest.	Successful self-governing entities have roles in monitoring and compliance. A nested framework would enable government to have ultimate responsibility for enforcement. (No guarantee that government could manage the resource better or achieve effective compliance without concurrence of the irrigators.)
Irrigation entity unwilling to take responsibility for co-management	Costs not worth benefits.	Entity is engaged by government in developing the 'rules' (in a water sharing plan) so that all parties' needs are accommodated. Incentives might include: more clearly defined entitlements; better understanding of reliability of supply (in over-used systems); tradeable water entitlements; and compensation for reduced entitlements (Ross and Martinez-Santos 2008)

Adapted from Baldwin *et al.* 2008

credibility gap (Joyce and Thomson 2000). Building 'reputation capital' is a significant business management strategy based on recognition that what a business does in any location and with any stakeholder, will contribute to the broader industry reputation: it can become a competitive advantage. Reputation capital predisposes communities to enter into open discussion rather than opposition, reducing the upfront costs and risk associated with gaining social acceptability. Ultimately, though, trust is needed for a social licence to be granted.

Trust relies on building social capital; it relies on:

> *'... collaborative capacity, the stock of active connections among people: the trust, mutual understanding and shared values and behaviours that bind the members of human networks and communities.' (Thomson and Joyce 2008)*

Trust is the willingness of both parties to be vulnerable to risk or loss through the actions of the other. A commitment to achieving industry standards and a contractual relationship between institutions can reduce social risk and enhance trust.

One of the underlying reasons that the state government did not agree to co-management in the Lockyer was based on a lack of trust. For many years, vociferous members of the irrigation community had been hostile and opposed to government regulation. For the most part, this had been to their private benefit and at a cost to other irrigators and the environment (public benefit). It had contributed to the creation of inequitable regulation and resource depletion. Building adequate trust would require the LWUF to be supported by all irrigators and be voicing words and taking action to manage the resource with credibility reflecting equitable treatment of members.

Local populations have a desire to have some measure of control over their own future. Increasing expectations need to be matched with scrutiny of actions. Social licence needs to be earned, then maintained. The increased irrigator responsibilities for sustainable water management under the NWI mean that community-based water resource management is appealing. It can provide support to irrigators and a unified entity for government to deal with. One of the challenges is to tailor community-based arrangements to suit the individual situation, particularly in self-supply areas such as the Lockyer where infrastructure has not provided irrigators with a common interest to date. Ultimately, building a strong social licence to irrigate through co-management will rest on the rules being transparently and collaboratively developed.

Acknowledgements

The authors acknowledge the support of the Department of Water, Western Australia for the research on which part of this chapter is based.

References and further reading

Baldwin C (2008) Integrating values and interests in water planning using a consensus-building approach. PhD thesis. University of Queensland, St Lucia.

Baldwin C, Hamstead M and Uhlmann V (2008) 'Interjurisdictional analysis of self-management governance arrangements for water resource management in Western Australia'. Report to Department of Water, Perth.

Bromley D and Cernea M (1989) 'The management of common property natural resources: some operational fallacies'. World Bank Discussion Paper No. 57, World Bank, Washington, DC.

Fisher R, Ury W and Patton B (1991) *Getting to Yes: Negotiating an Agreement without Giving In*. 2nd edn. Random House, Sydney.

Joyce S and Thomson I (2000) Earning a social licence to operate: social acceptability and resource development in Latin America. *The Canadian Mining and Metallurgical Bulletin* **93**, 1037.

LWUF (Lockyer Water Users Forum) (2006) 'General points of agreement between irrigators regarding the water resource plan'. Unpublished report of meeting, 5 December 2006, Gatton.

McCay BJ and Jentoft S (1996) From the bottom up: participatory issues in fisheries management. *Society and Natural Resources* **9**, 237–250.

McKean M (1992) Management of traditional common lands (Iriaichi) in Japan. In: *Making the Commons Work: Theory, Practice and Policy*. (Ed. D Bromley) pp. 63–98. Institute for Contemporary Studies Press, San Francisco.

NRM (Department of Natural Resources and Mines) (2005) 'Moreton draft water resource plan: information report'. Department of Natural Resources, Mines, Brisbane.

Ostrom E (1990) *Governing the Commons: The Evolution of Institutions for Collective Action*. Cambridge University Press, New York.

Ostrom E (2005) *Understanding Institutional Diversity*. Princeton University Press, New Jersey.

Ostrom E, Burger J, Field CB, Norgaard RB and Policansky D (1999) Revisiting the commons: local lessons, global challenges. *Science* **284**(5412), 278–282.

Ostrom E and Nagendra H (2007) Tenure alone is not sufficient: monitoring is essential. *Environmental Economics and Policy Studies* **8**(3), 178–199.

Ross A and Martinez-Santos P (2008) 'The challenge of collaborative groundwater governance: four case studies from Spain and Australia' [cited 12 October 2008]. Available from <http://www.newater.uos.de/caiwa/data/papers%20session/F4/ARPMSCAIWA.pdf>.

Sarker A, Baldwin C and Ross H (2009) Managing groundwater as a common-pool resource: an Australian case study. *Water Policy* **11**(5), 598–614.

Thomson I and Joyce S (2008) 'The social licence to operate: what it is and why does it seem so difficult to obtain?' Presentation to Prospectors and Developers Association of Canada (PDAC) Convention, Toronto, March 2008.

Weinstein M (2000) Pieces of the puzzle: solutions for community-based fisheries management from native Canadians, Japanese cooperatives, and common property researchers. *Georgetown International Environmental Law Review* **12**(2), 375–406.

16

A conceptual framework for sustainable agriculture

Jacqueline Williams

This chapter presents a conceptual framework to reconnect the linkage between urban and rural communities, sustainable food production and healthy ecosystems. The framework provides a means to contribute to the *social licence to farm* through the streamlining and harmonisation of reporting requirements and the creation of a recognition system for sustainable food and fibre products and equitable cost sharing of public good ecosystem services. The conceptual framework is not presented as the definitive design. Rather it is intended as a stimulus to further development of a unified national approach, which harmonises and streamlines key elements.

The research presented in this chapter was developed through a collaborative case study with Macintyre Brook irrigators in Queensland. This led to the development of a framework subsequently tested through national and state-based focus groups. This research was undertaken as part of the CRC for Irrigation Futures System Harmonisation Program. The chapter outlines the issues involved in the creation of an integrated framework for evaluation and formal recognition of the sustainability outcomes of farming. It then considers how such a system might be created by harmonising existing public and private sector arrangements. Finally, the chapter reports some preliminary responses to the conceptual framework from discussion groups. Together, these elements suggest that it is possible to create a more comprehensive system to objectively monitor (and reward) farm sustainability.

The social brand dilemma

Demonstrating sustainability of the Australian agricultural sector has become a core challenge for farmers attempting to secure a social licence. Differing perceptions of sustainability have resulted in numerous strategies by government, the market and community. These range from regulatory plans, quality assurance programs, environmental management systems (EMSs), property management planning processes and value chain systems. Dealing with this plethora of systems has become a serious issue for many farmers in Australia who are attempting to demonstrate their sustainability to government, markets and the community in the absence of an integrated, recognised system (Martin 1991; Martin and Verbeek 2000; Williams *et al.* 2008; Australian Government 2008a; Productivity Commission 2007). This burden is well illustrated by the expenses incurred by family farms, with 'red tape' now accounting for 14% of net farm profit (Productivity Commission 2007). Overcoming this impediment to demonstrating sustainability requires two reforms: first the harmonisation of the 'architecture' of environmental and other laws (Martin *et al.* 2007), and second institutional harmonisation to deal with the duplication and complexities created by the government–market–community nexus around sustainable agriculture (Martin *et al.* 2008). This need was further emphasised in the Australian 2020 Summit, which recommended the harmonisation of regulations as a priority for rural and regional Australia (Australian Government 2008a). This chapter will, however, focus solely on the second reform of institutional harmonisation, because regulatory harmonisation is discussed elsewhere.

Reviews of environmental assurance schemes intended to support social licence highlights many challenges. These include the failure of EMSs and other environmental assurance programs in the market place (Grains Council of Australia 2006; Pahl 2007; Pahl and Sharp 2007; Bhaskaran *et al.* 2007; Keogh 2006; Chang and Christiansen 2006; Gunningham 2007), and the confusion, duplication, increased management and industry costs of quality assurance (QA) and EMS, which provide limited assurances to customers and the community about food safety, animal welfare and environmental stewardship (Richards *et al.* 2007). Coupled with these domestic drivers are international pressures to demonstrate sustainability, driven mainly through major retailers as a demonstration of their commitment to corporate social responsibility. These corporate requirements, integrated within the farm supply chain, create another layer of assurance farmers have to meet on top of regulatory and community requirements. This dilemma is not unique to Australia, with other systems starting to appear in markets such as the US with the development of the *Sustainable Agriculture Practice Standard for Food, Fiber and Biofuel Crop Producers and Agricultural Product Handlers and Processors (SCS-001)* (Leonardo Academy 2010). The European Commission (2005) recognising the need for harmonisation, undertook major reform of their *Common Agricultural Policy* (CAP) in 2003 with an emphasis on cross-compliance. Cross-compliance requires farmers to demonstrate that land is kept in *Good Agricultural and Environmental Condition* (GAEC) and compliance with *Statutory Management Requirements* (SMR) relating to: environment, public/ animal/plant health and animal welfare. Cross-compliance must be undertaken by EC farmers to receive government assistance through the Single Farm Payment Scheme (SPS). Further international pressures for Australian farmers are evident, with the report *Agriculture at a Crossroads: Global Report* by the International Assessment of Agricultural Science and Technology for Development (2009) recommending that Australia adopt more socially and environmentally sustainable agriculture, with a focus on agro-ecological approaches.

How can a farmer help to secure the sector's social licence through demonstrating sustainability when Australia lacks a nationally consistent farm-to-nation system to demonstrate the results of positive efforts by farmers, government, industry and community? This information vacuum contributes to the ongoing market failure of adequate recognition and trading of more sustainable food, fibre and ecosystem services.

Farmers have also identified the lack of recognition of on-farm collected data within the natural resources management system as contributing to the loss of aggregated farm-level natural resource management data for government decision-making (Martin *et al.* 2008). Although there are arguments about the quality of on-farm collected data, the lack of use of 'grass roots' information can lead to perverse outcomes, particularly a failure to readily detect when what is contained in a scientific model or report is blatantly inconsistent with the evidence of local observers (Searle *et al.* 2005). Responsibilities for resource condition monitoring are unclear. Regional Natural Resource Management (NRM) bodies charged with this responsibility were found to be unsure of what data were required, where the gaps were, or how to report against resource condition targets (Land and Water Australia 2005).

The *Australia State of the Environment 2006* report identified that a consistent set of indicators for assessing the state of natural resources does not exist in Australia (Beeton *et al.* 2006).

This creates difficulties in determining the state of the natural resource base. Important potential benefits such as carbon sequestration and water quality improvement, which have been called for by numerous bodies (National Farmers Federation and Australian Conservation Foundation 2000; Business Leaders Roundtable 2001) may actually be occurring in the landscape, but go unrecognised and unrewarded. Although new business models for farmland conservation are potentially attractive, the absence of suitable institutional arrangements for measurement and reporting is retarding their development (Martin *et al.* 2007).

With 60% of Australia managed for agricultural land-use, farmers are by far the largest investors of NRM, investing $3 billion NRM annually (ABS 2006; ABS 2008) and $314 million annually in water-related management activities (ABS 2007). Analysis of Regional NRM bodies on-ground investments (Williams 2007a) compared with the ABS on-farm NRM investment (ABS 2006) indicates missed opportunities in recognising sustainable agriculture. The majority of on-farm NRM investments by farmers are overlooked in regional NRM reporting systems. Trading opportunities that could improve cost-sharing arrangements for public good outcomes from agriculture are unlikely to develop unless such problems are overcome.

These issues have been recognised by Australian Ministerial Councils for Natural Resources and Primary Industries, with a working group established in 2006 to tackle the issues through the development of a 'National Framework for Farm Management Systems'. However, negotiations between the states reached an impasse and the framework never came to fruition. *The Signposts for Australian Agriculture Program* (Bureau of Rural Sciences 2005) proposed a national approach to triple bottom line reporting for the agricultural industries, but this work was not translated to the farm scale. Recent rationalisation of the Department of Agriculture, Forestry and Fisheries has resulted in it having a lesser status. Despite numerous Government directives, such as those from the Ministerial Councils and the recommendations from the 2020 Summit, little progress has occurred.

Overcoming these challenges could be achieved through a national approach to farm environmental policy using a range of incentives to reward farmers for delivery of public good outcomes, coupled with a national assurance scheme (Pahl 2007) or through the creation of a more robust private market, supported by taxation reform (Martin and Werren 2009). Developing a robust national approach focused on sustainable agriculture requires consistent standards and indicators, underpinned by government recognition and endorsement (Williams 2007a, 2007b). The institutional challenges for such an approach are substantial, but the success of the cotton industry in achieving a limited form of co-regulation in Queensland holds promise (Martin *et al.* 2007). A sustainable agriculture framework could provide the opportunity to harmonise regulatory requirements (such as property management plans) and institutional elements. The benefits would lower transaction costs, better protect the social licence to farm, provide a focus for on-farm data collection linked to regional and national outcomes, and enable trade-offs and rewards for better performance through a recognised system.

The CRC Irrigation Futures' *System Harmonisation* research with Macintyre Brook irrigators in South-West Queensland provided an opportunity to explore in detail, in partnership with primary producers, the potential for an integrated regional approach to assessing, reporting and (hopefully) rewarding sustainability outcomes from improved irrigation and land management.

Case study: Macintyre Brook catchment

Macintyre Brook is a sub-catchment of the Murray–Darling Basin in the NSW/Queensland Border Rivers region. The Macintyre Brook sub-catchment covers 4500 km^2, with 2050 hectares of this area irrigated, the balance consisting of natural vegetation, production forestry and dryland farming. The irrigated farmland in the Macintyre Brook produces lucerne, citrus, stone fruit, vines, olives, cereals and pastures for stock grazing (Macintyre Brook Irrigators Association 2010).

As a sub-catchment of the Border Rivers, Macintyre Brook falls under the complex governance of the Murray–Darling Basin. Resource management issues are compounded by a lack of baseline data. Of particular concern has been the lack of soils data and knowledge of the interaction of the surface and groundwater hydrology. This makes resource assessment and on-farm and catchment-wide management difficult.

Macintyre Brook irrigators' issues mirror broader issues of NRM across Australia. These issues include: the lack of reliable environmental performance or environmental status data for farming enterprises; difficulties in data access and exchange; problems arising from modelling (and competing models); and the insufficiency of signals to the market place and government of the performance of farming operations. These problems make it difficult to assess environmental performance objectively and to develop statutory NRM plans. To confront these issues requires a framework that streamlines the requirements of Property Management Systems (PMS), statutory Land and Water Management Plans (LWMP) and related property management and administration. Ideally, such a system would allow the best environmentally performing producers to improve consumer information to further stimulate environmental performance, with better monitoring and communication of performance to markets and regulators.

A recognised EMS standard with strong support from Australian Governments has been identified as urgently needed in Australian agriculture (Keogh 2006). Chang and Kristiansen (2006) highlight the need for concrete proof of environmental and quality credentials to enable effective competition in the export market. Gunningham (2007) proposed a strategy that builds on the strengths of a voluntary EMS combined with complementary policy instruments such as appropriate performance indicators, accountability, transparency and consultation, coupled with economic incentives and regulatory underpinnings to form a financially attractive mix.

The development of the conceptual framework

The aim of the research was to develop a conceptual framework to support the social licence to farm through a marketable claim of superior sustainability, backed by credible recognition of environmental performance, and by communication and training. Such a system requires improvements in data sharing, coordination among agencies and adjustment to regulations or institutional arrangements. The goal was to test potential local, state and national level solutions.

Drawing on such proposals, Macintyre Brook irrigators sought a recognised system at the farm scale to: enable trading of ecosystem services; demonstrate duty of care compliance at the farm scale; pilot a framework for Queensland that aligns with existing state, national and industry targets, standards and indicators; develop a system to identify and purchase stewardship through a pro-active property management system recognised at the regional level, and data sharing through consistent standards, indicators and systems that feed into regional natural resource management and state of the environment (SoE) reporting.

The main barriers to gain an improved social licence based on better sustainability performance that were identified in Macintyre Brook were:

- inconsistent processes and language between farm-level statutory plans, leading to duplication of effort and confusion
- the variety of regulatory and non-regulatory standards, plans and approvals, which restrict the ability of markets to objectively evaluate farmers' sustainability performance

- the lack of baseline data and natural resource monitoring. Natural resource data collection is undertaken by a variety of organisations without mechanisms to standardise, validate and share the data to ensure its utility for multiple users.

The research resulted in a framework based on recognition to support the trading of sustainable agriculture products, such as sustainable food, fibre and ecosystem services. The framework was developed collaboratively with the local irrigators, with consultation with state and federal government personnel. The framework (Figure 16.1) is underpinned by:

- a property management planning system (PMPS) harmonisation encompassed within a standard for sustainable agriculture
- certifying bodies for sustainable agriculture product labelling and verification
- the use of existing policy instruments and institutional arrangements that can be adapted and adopted for this purpose.

The existing policy instruments and institutional arrangements encompassed in the framework include: the *Natural Heritage Trust of Australia Act 1997* (Cwlth); *Income Tax Assessment Act 1997* (Cwlth); National Framework for NRM Standards, Targets, Monitoring and Evaluation (Natural Resource Management Ministerial Council 2002); SCARM Sustainable Agriculture indicators (SCARM 1998); National Framework for NRM National Coordination Committees on Resource Conditions (Australian Government 2008b); Intergovernmental Agreement on the Environment 1992 (Commonwealth of Australia 1992), and the Queensland NRW Land and Water Management Plan Recognition Framework (Queensland Department of Natural Resources and Water 2008a). The framework also incorporates other instruments identified in prior research (Martin *et al.* 2007; Williams 2007a) and lessons from the evolution of the organic standard in Australia (Organic Federation of Australia 2008; Standards Australia 2010).

Elements of the framework

The framework was developed based on existing institutions with a focus on streamlining, harmonisation and coordination. The framework incorporates: property management planning systems (PMPS); national accreditation; a sustainable agriculture standard; and pathways to recognise sustainable food and fibre and invest in ecosystem services. The conceptual design is based around developing and recognising the PMPS as the instrument for certifying good standards of environmental management. The following discussion will provide an overview of the structure and pathways of the framework.

There are numerous property management planning systems that farmers are required to adopt for regulatory NRM, NRM grants, quality assurance and other market requirements including industry programs. Coupled with these are the nationally recognised sustainable agriculture indicators (SCARM 1998), used by the Australian Bureau of Statistics to track the status of sustainable agriculture in Australia. Earlier work (Williams 2007b) found that these often duplicated and inconsistent approaches could be streamlined and harmonised down to common elements under the themes of:

- *environment:*
  - resource assessment, including baseline natural resource assessment, conditional natural resource assessment and land suitability elements
  - farm action planning using environmental risk assessment and management
- *social:*
  - human safety, including human risk assessment and management (food safety and farm safety)
  - animal welfare, including animal risk assessment and management (welfare and ethics)

- *economic:*
  - economic and social sustainability (sustainability evaluation).

Under each of these themes are common elements or modules of PMPS potentially aligned to nationally agreed methodologies and indicators and examples of best management practice and recommended guides. PMPS can provide a means to synthesise statutory NRM plans, industry, market and community property plans into a 'one stop shop' approach to avoid duplication and ensure that the measures and methodologies used are consistent. This would make it possible to aggregate data from the farm level to nation level. What is missing to enable this 'one stop shop' approach to property planning is recognition by government (Ecker *et al.* 2005).

Mechanisms available to government to accredit PMPS

There are a number of pathways that governments can use to support objective accreditation of PMPS. The National Framework for NRM Standards, Targets, Monitoring and Evaluation (Natural Resource Management Ministerial Council 2002) and the National Framework for NRM National Coordination Committees on Resource Condition (Australian Government 2008b) have established agreed resource conditions, indicators and methodologies that can be translated to farm scale via the PMPS. The *Natural Heritage Trust of Australia Act 2007* (Cwlth) s 16 provides a legal definition of 'sustainable agriculture', clearly identifying that a property management plan can be used as a means to demonstrate sustainability:

> '(1) *For the purposes of this Act, sustainable agriculture means the use of agricultural practices and systems that maintain or improve the following:*
> *(a) The economic viability of agricultural production;*
> *(b) The social viability and well-being of rural communities;*
> *(c) The ecologically sustainable use of Australia's biodiversity;*
> *(d) The natural resource base;*
> *(e) Ecosystems that are influenced by agricultural activities.*
>
> (2) *To avoid doubt, for the purposes of this Act, property management planning in relation to the farm unit is taken to be sustainable agriculture.*'

Further, The Intergovernmental Agreement on the Environment (Commonwealth of Australia 1992) provides mechanisms for accreditation and removing duplication between the Commonwealth and the states in Section 2.5 Accommodation of Interests. These existing institutional mechanisms could provide the means for the Australian Government to facilitate proper recognition of a PMPS, in partnership with the states and industry. The Queensland Government's recognition framework for Land and Water Management Plans (Queensland Department of Natural Resources and Water 2008) and the 2005 Memorandum of Understanding between the Queensland Government and Queensland Farmers Federation (Queensland Government 2005) provide good illustrations of how such a recognition framework can be developed.

Through the PMPS process, a better pathway for NRM investment may be created. Because the farm-scale data has been collected using nationally consistent indicators and methodologies, this data can be aggregated to inform the nation of the condition of the resource base, potential or realised improvements in the landscape, and the sustainability of rural communities. An opportunity exists to create a mechanism for data exchange with farmers potentially attracting a rebate through the Australian Taxation Office. The 'Landcare Operations' provision under the *Income Tax Assessment Act 1997* (Cwlth) provides for this concession. It is underpinned by the requirement for an approved land management plan and there is a register of authorised consultants who can prepare such a plan. This system could be adapted to facilitate farm-scale data collection through the PMPS. Regional NRM bodies and Landcare groups (if recognised as authorised consultants) could assist by providing technical expertise and certify the PMPS undertaken by the farmer.

The PMPS approach also provides a way for farmers to demonstrate and trade ecosystem services in the market place, because PMPS could provide proof of a scientifically valid baseline assessment and proof of delivery of ecosystem services through monitoring. Martin *et al.* (2007) have proposed a regional conservation business model for the trading of ecosystem services consisting of a Farmland Conservation Trust, which would run a farmland conservation philanthropic fund, a farmland conservation research fund and a farmland conservation investment scheme. The trust would be underpinned by a regional collective approach to natural resource management for the delivery of ecosystem services to investors of the trust. Martin *et al.* (2007) found that including the management of ecosystem services in the taxation definition of 'primary production' and the establishment of professionally managed conservation zones for ecosystem services as 'forestry' will increase the investment in farmland conservation and provide more effective cost sharing of public good farmland conservation.

Sustainable agriculture standard

The development of an accredited PMPS could also provide a direct pathway for the development of a certification standard for sustainable agriculture. A consumer-credible certification scheme for farm produce is seen by many as one way of helping to provide some market reward for sustainable farming. Lessons from the evolution of the Australian Standard for Organic and Biodynamic Products and Regulation (Standards Australia 2010; Organic Federation of Australia 2008) demonstrate how such a standard could be developed.

The new Australian Standard AS 6000-2009 Organic and biodynamic product (2009) sets out the minimum requirements to be met by growers and manufacturers operating in the organic and biodynamic industry. This remedies an issue that had concerned the organic produce industry for a number of years: the need for a unified national standard. The issue had arisen because of the existence of numerous standards and definitions for organic and biodynamic. An Australian Federal Court decision in 2007 found that the absence of a legal definition of the term 'organic' made it difficult to create an enforceable injunction against misrepresentation of 'organic' in the market place. This finding led to the development of a uniform, accepted standard for organic and biodynamic produce.

The new standard was developed with input from stakeholder groups, including the industry, consumers, retailers and regulators. The organic and biodynamic standard is a co-regulatory system and is such voluntary. However, the Australian Competition and Consumer Commission (ACCC) considers the standard as assisting their enforcement activities under the *Trade Practices Act 1974* (Cwlth) against misleading and deceptive conduct and false and misleading representations. The Australian standard for organic and biodynamic produce has provided the means to ensure the integrity of products that are sold as organic and biodynamic (Organic Federation of Australia 2008).

Responses to the conceptual framework

To test the feasibility of the framework, focus groups were undertaken in 2008. These included participants from industry, government (Australian and state), research institutes, regional NRM bodies and the farming community. The focus groups explored the institutional supports and impediments to the framework. The focus groups were based on the following key questions:

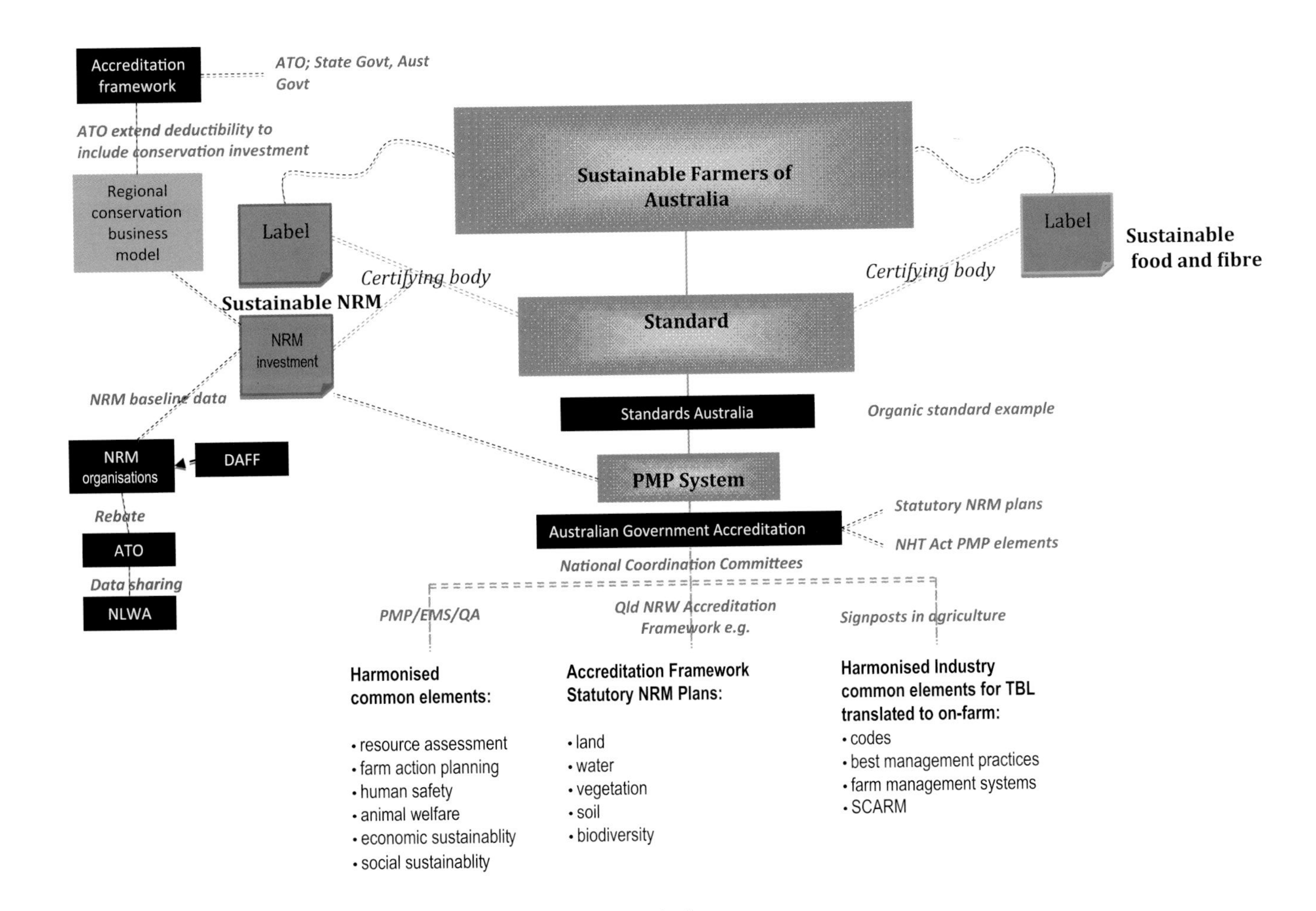

Figure 16.1: The elements of a conceptual framework for sustainable agriculture

- Does the framework fulfil sustainable agriculture policy objectives?
- Does the framework support fair trade of all goods and services?
- What are the required roles and responsibilities to enable the framework to work?
- What institutional arrangements are required to enable the framework?
- What are the risks?

The following discussion summarises the focus groups' feedback to the proposed framework. Because of the nature of the feedback, the views are presented as a simple narrative, highlighting matters raised by the participants in the discussion. It is an indication of issues relevant to the different stakeholder groups.

Although farm EMS has been lauded by many as a means to demonstrate sustainability in agriculture, it was considered by the focus groups to be inadequate because of the lack of NRM baseline data integrated in the process. They felt that farmers need a system that clearly demonstrates sustainability in response to community pressure, government regulations and the 'market' demands. Farm EMS are perceived not to deliver demonstrated value beyond the farm, nor to connect with regional/state/national and international arenas. It was noted that the national Land and Water Audit failed to recognise ecosystem and environmental services, despite farmers collecting the relevant data and providing the potentially valuable environmental services. In addition, there was no mechanism for collection of, nor repository for, on-farm data. It was suggested that governments fail to recognise delivery of ecosystem services unless they have been formally associated with a government-funded program such as Landcare.

Existing industry programs, such as Cotton Australia's Best Management Practice (BMP) and the Australian Rice Industry's Environmental Champions Program, were seen as potentially good reporting systems, but were perceived to have failed to deliver the market premiums expected. This may possibly be associated with the lack of 'hard' data (for example, hectares under no tillage, kilometres of open channel replaced with pipes) as opposed to the currently collected 'soft' data, which only records what percentage of farmers apply any particular practice.

It was felt that government's return on investment in NRM could be better demonstrated if baseline and regular change of condition data were collected. Food security and declines in rural communities are of high priority for farmers, but water scarcity and emissions trading schemes are of greatest interest to the government and the media. Achieving and rewarding sustainable agriculture receives little attention.

The big strategic issues of sustainable agriculture, NRM, the 'right to farm' and management of biodiversity were considered by government participants as 'sleeper' issues that farmers will need to deal with in the not too distant future. It was agreed by the focus group participants that there was a need to elevate sustainable agriculture and the need for an effective integrated framework as a high priority for the government. For the farmer, the main driver is to streamline the costly requirements of state and national regulators, and to also demonstrate environmental stewardship to an increasingly interested export market.

It was suggested that any such integrated approach would have to be subject to a cost–benefit analysis because there may be risks. It was agreed that if the Australian Government recognised a national 'standard' approach, then the costs of compliance could be reduced. Active participation of all stakeholders in the development of the framework was needed. It was suggested that all levels of government need to give the framework development priority, with the Australian Government providing leadership. The state governments' role in achieving the framework implementation would be through the harmonisation/streamlining of regulatory aspects of NRM. It was suggested that the standard for sustainable agriculture measurements and indicators are already agreed; it is 'recognition' that is missing. Existing

NRM processes and platforms have the capacity to enable harmonisation. Bilateral agreements in place between different jurisdictions could drive creation and operation of the framework. Programs such as Caring for our Country could use PMPS approaches to assess the condition of the landscape, which could better inform further investment and assist in demonstrating sustainability outcomes. One matter of concern was whether PMPS assessments ought be carried out by land managers themselves or conducted by independent experts. It was pointed out that government agencies often preferred to recognise assessments carried out by other approved agencies. It was suggested that, due to limited coverage, such an approach gave an end result of only 20% accuracy. It was thought that simply because of the far greater coverage that can be achieved by farmers carrying out site assessments, greater reliability is likely through using land manager assessment.

The lack and quality of data was considered a big issue. It was highlighted that 'standards' have been developed by National Coordination Committees (Australian Government 2008b). Despite this, the necessary data (at all levels) were not being collected in compatible forms across jurisdictions. The benefit of the proposed standard approach was that once standards are recognised through the framework they could drive data collection, better informing national databases.

Any system needs to consider the varying scales at which property planning and monitoring information is collected and stored, and what is required for natural resource management and for sustainable farm management. A clear and agreed definition of 'sustainability' is required. It was noted that currently producers could not demonstrate their systems/practices were sustainable. This is a major driver of interest in such a proposed framework from the farming sector. Government expressed an interest in the potential of the framework to support reporting on the Emissions Trading Scheme, sustainable NRM and land use adjustment in catchments (connected also with National Land and Water Audit).

Conclusion

An agreed process involving the Australian and state governments in partnership with industry is required to create a unified framework for recognition of sustainable agriculture. Such a framework would be an important factor in preserving the social licence to farm, by providing greater transparency about ecological outcomes achieved by farmers. This research suggests this must be guided by using the existing recognised standards as an agreed 'language' of sustainable processes to give a consistent approach to measurement. Adoption of objective outcome reporting will be greatly advanced if it is tied to market or taxation incentives, and these are possible to at least some degree. The 150 000 small farm businesses and 500 000 peri-urban farms in Australia need to be engaged in the process to ensure the framework is owned and supported by primary producers.

A framework such as the one proposed offers a pathway for institutional harmonisation, which provides a means to support farmers' social licence through recognition of sustainable food and fibre production and greater awareness of the provision of public good ecosystem services from farm management practices. It is clear from the research, including detailed work with the irrigators of Macintyre Brook and discussions at all levels of government (including a number of states), that a harmonised recognition scheme ought be possible. Such a scheme, if reasonably comprehensive and objective, can help reduce impediments to markets and guide governments to better reward superior sustainability performance by farmers. It can assist in targeting investment, and it can assist in communicating investment outcomes. The current institutional arrangements are complex and unduly burdensome. What is needed is a powerful political impetus to create such an integrated and streamlined approach.

References and further reading

ABS (Australian Bureau of Statistics) (2006) 'Natural resource management on Australian farms 2004–05'. Australian Bureau of Statistics, Canberra.

ABS (2007) '4613.0 – Australia's environment: issues and trends'. Australian Bureau of Statistics, Canberra.

ABS (2008) '4620.0 – Natural resource management on Australian farms, 2006–07'. Australian Bureau of Statistics, Canberra.

Australian Government (2008a) 'Australian 2020 summit final report: future directions for rural industries and rural communities'. <http://www.australia2020.gov.au/final_report/index.cfm>.

Australian Government (2008b) 'National framework for NRM national coordination committees on resource condition'. Australian Government, Canberra.

Beeton RJS, Buckley K, Jones GJ, Morgan D, Reichelt RE and Trewin D (2006) 'Australia state of the environment 2006'. Independent report to the Australian Government Minister for the Environment and Heritage, Canberra.

Bhaskaran S, Polonsky M, Cary J and Ferdandez S (2007) Environmentally sustainable food production and marketing opportunity or hype? *British Food Journal* **108**(8), 677–690.

Bureau of Rural Sciences (2005) 'Signposts of agriculture: preliminary framework and collation of industry profiles'. Bureau of Rural Sciences, Canberra.

Business Leaders Roundtable (2001) 'Repairing the country: leveraging private investment'. Allens Consulting Group, Sydney.

Chang C and Kristiansen P (2006) Selling Australia as 'clean and green'. *The Australian Journal of Agricultural and Resource Economics* **50**, 103–113.

Commonwealth of Australia (1992) 'Intergovernmental agreement on the environment'. Department of Environment, Water, Heritage and the Arts, Canberra.

Ecker S, Williams J and Kinnimonth I (2005) 'ROOFS concept paper and implementation plan'. Ecker Consulting, Canberra.

European Commission (2005) 'Cross compliance'. Directorate-General for Agriculture and Rural Development of the European Commission, Brussels, <http://europa.eu.int/comm/agriculture/index_en.htm>.

Grains Council of Australia (2006) 'Market signals for environmental and other assurances in the supply chain'. Grains Council of Australia, Canberra.

Gunningham N (2007) Incentives to improve farm management: EMS, supply-chains and civil society. *Journal of Environmental Management* **82**, 302–310.

International Assessment of Agricultural Knowledge, Science and Technology for Development (2009) *Agriculture at a Crossroads: The Global Report*. Island Press, Washington.

Keogh M (2006) Editorial. *Farm Policy Journal* **3**(4), 4–5.

Land and Water Australia (2005) 'National monitoring and evaluation symposium'. Australian Government, Hobart.

Leonardo Academy (2010) 'Sustainable agriculture practice standard'. Leonardo Academy, Madison, <http://www.leonardoacademy.org/programs/program-projects/182-sustainable-agriculture-practice-standard.html>.

Macintyre Brook Irrigators Association (2010) Macintyre Brook Irrigators Association, Inglewood, <http://www.macbrook.com.au>.

Martin PV (1991) Why Australia fails to exploit publicly funded R&D. *Prometheus* **9**(2), 362–378.

Martin P and Verbeek M (2000) 'Cartography for environmental law: finding new paths to effective resource use regulation'. Land and Water Research and Development Corporation, Canberra.

Martin P and Werren K (2009) The use of taxation incentives to create new eco-service markets. In: *9th Global Conference on Environmental Taxation: Critical Issues in Environmental Taxation*. 6–7 November. Asia-Pacific Centre for Environmental Law, of the Faculty of Law, National University of Singapore, later published in 7th edn of *Critical Issues in Environmental Taxation*, Oxford University Press 2009.

Martin P, Bartel RL, Sinden JA, Gunningham N and Hannam I (2007) 'Developing a good regulatory practice model for environmental regulations impacting on farmers'. Land and Water Australia, Canberra.

Martin PV, Williams JA and Stone C (2008) 'Transaction costs and water reform: the devils hiding in the details'. CRC for Irrigation Futures Technical Report 08/08, Richmond.

National Farmers Federation and Australian Conservation Foundation (2000) 'National investments in rural landscapes'. Virtual Consulting Group & Griffin nrm Pty Ltd, Canberra.

Natural Resource Management Ministerial Council (2002) 'National framework for natural resource management standards and targets'. Australian Government, Canberra.

Organic Federation of Australia (2008) 'OFA position paper: the Australian Standard for Organic and Biodynamic Products and Regulation'. OFA, Mossman.

Pahl LI (2007) Adoption of environmental assurance in pastoral industry supply chains – market failure and beyond. *Australian Journal of Experimental Agriculture* **47**, 233–244.

Pahl LI and Sharp R (2007) Stakeholder expectations for environmental assurance in agriculture: Lessons from the pastoral industry. *Australian Journal of Experimental Agriculture* **47**, 260–272.

Productivity Commission (2007) 'Annual review of regulatory burdens on business'. Productivity Commission Draft Research Report. Commonwealth of Australia, Melbourne.

Queensland Department of Natural Resources and Water (2008) 'Land and water management plan recognition framework'. Queensland Government, Brisbane.

Queensland Government (2005) 'Memorandum of understanding'. Queensland Government, <http://www.dpi.qld.gov.au/4789_9620.htm>.

Richards CL, Morgan CK and Baldwin BJ (2007) Accreditation in Australian agriculture – on the right track or are farmers lost in the maze? *AFBM* **3**(1), 31–38.

SCARM (1998) 'Sustainable agriculture: assessing Australia's recent performance'. Standing Committee on Agriculture and Resource Management, Canberra.

Searle R, Oliver P, Griffith G, Salloway M and Weber T (2005) Implementation of a water quality model by the community – some lessons. *Proceedings MODSIM, International Congress on Modeling and Simulation*. Modelling and Simulation Society of Australia and New Zealand, Canberra.

Standards Australia (2010) Organic and biodynamic products: the new regime. *Standards Australia E-newsletter* Issue 1, July 2010.

Williams J (2007a) 'Success attributes of regional NRM systems in Australia'. PhD thesis. University of Queensland, St Lucia.

Williams J (2007b) 'Queensland property management systems common elements'. Draft Report. AgLaw Centre, University of New England, Armidale.

Williams JA, Sauer I, Love C, English G, Jackson M, Cherry J, Sansom G, Wood B, Russell A, Williams J, O'Sullivan J and Ecker S (2008) 'A national framework for sustainable agriculture'. August 2007, DAFF Canberra Workshop Proceedings, Canberra.

FUTURE DIRECTIONS

17

Renegotiating farmers' social licence

Paul Martin and Jacqueline Williams

In this final chapter, we bring together many of the strands from the previous chapters. We highlight some themes from the experience of farmers and farmer organisations, in part viewed through the lens of the more academic chapters. We suggest some potential directions for more effectively tackling social licence issues in farming. Of all the lessons that emerge, one stands out: renegotiating the social licence of farmers will require committed leaders and cohesive farmer organisations able to secure the trust of an often-sceptical public. This requires that they embrace unpalatable criticisms so that they can develop strategies to deal with them to the extent that is possible. Such adjustments are often likely to require significant changes to historically established farming practices, and new governance structures in the farm sector. Only within the context of a sophisticated strategy of genuine engagement with the demands of the community is it likely that tactical interventions, such as public relations, voluntary impact reporting and political lobbying, will prove to be effective in the longer term in defending the maximum freedoms for farming. If it is possible to create a genuine partnership with government and non-government organisations outside the farm sector, then co-regulation that will better reward greater social responsibility is possible. If not, then farmers may find themselves continually on the back foot, responding defensively to increasingly costly and complex regulatory requirements.

Democracy, the law and social licence

Few who argue in support of farmers' rights would suggest that democracy is a defective system of government, or that national constitutions ought be flaunted to benefit particular interests. Indeed, frequently arguments to protect the freedom of farmers to pursue their vocation and enjoy their property unhindered are based upon the argument that constraints on farmers violate principles of democracy or the rule of law. However, democracy entrenches the right of the majority to constrain the actions of the minority, even if the affected minority is unhappy with the result, and this applies to farmers no less than to any other group in society. Chapter 1 highlights that farmers' rights, like those of every citizen, are bounded by what society is prepared to support and defend. This is an ever-changing bargain between the citizen and the state. Farmers' social licence is not 'etched in stone': it is constantly adjusted through a sort of negotiation with the community about what the community will provide as privileges and supports for farming, and what it will impose as restrictions on farming.

Of course, the capacity to restrict the freedom of the few to advance the interests of the majority is not unconstrained. Generally, the constitutions of democracies place restrictions on 'the tyranny of the majority'. Constitutional provisions require compensation for government acquisition of private property, but in most democracies the conditions under which compensation is payable are closely restricted. Even in the USA, where property and human rights are so deeply intertwined in the political psyche, government has great latitude to restrict uses to which private property can be put without having to pay compensation for claimed expropriation (Martin 2010). This makes sense, because otherwise the right of the majority to make laws to control what happens within the democracy would be economically impractical to exercise, because many politically significant restrictions would require payment from the public purse. Democracy would not be a system governed by the will of the people, exercised through parliament – we would instead have a form of market place for rights and freedoms, which would be quite different from Parliamentary democracy and the rule of law as most people would understand it to be.

It is for such reasons that farmers' property rights cannot be absolute and unchanging, and can always be adjusted by laws, though government is constrained to some degree from appropriating farmers' property without compensation (Martin 2010).

The politics of social licence

The democratic system of legal rights and obligations, freedoms and restrictions involve a complex dynamic through which political choices are made. Citizens decide their preferences and pursue them individually or through non-government organisations that lobby politicians and mount political campaigns. The media plays a significant part in the contest of ideas and interests, highlighting conflicts and exposing the actions of all 'players' in the system to scrutiny. The press and various experts, commentators, and professional lobbyists and communicators shape opinions in ways that support political positions, for philosophical, intellectual or merely commercial reasons.

It is in this setting that the social licence for farming is determined. From the preceding chapters, we have seen that the messy processes of democracy can work in favour of farmers' interests, or against them. The farm-friendly USA has shown a shift first to increase support for farmers by restricting the legal rights of other citizens to 'interfere' with farming, but more recently has imposed upon farmers increasing restrictions, and reduced the scope of the special protections from civil actions that they have enjoyed under 'right to farm' laws. Several chapters have shown a wide variety of approaches in Australia, the USA, China, India, Bhutan, Europe and Iceland to reshaping farming to meet the expectations of the majority, interpreted through political processes. The mechanisms used in all these jurisdictions is a mixture of regulatory sticks and fiscal carrots, but all demonstrate the general trend towards greater requirements for sustainable farming practices and protection of the health of the general community from potential adverse effects of (particularly) industrial farming practices.

One of the drivers of the change in political, legal and economic support for farming has been the success of industrial farming itself. As industrial farming has become a large-scale, low employment and highly profitable activity, with a reducing interaction between the industrial farm and the 'host' communities for their activities, the public good reasons to provide it special support have diminished. Industrialisation also concentrates potentially offensive or harmful by-products of farming, such as runoff, chemicals or intensity of animal husbandry. Another driver of changed attitudes towards farming has been the growth of other industries that have dwarfed the economic and employment contribution of farming, at the same time as community values have shifted towards demanding more protection for the environment, with increasing demands for 'virtue' in the use of resources, exploitation of animals and protection of workers. A plethora of media exposé programs that (rightly or wrongly) criticise the heavily industrialised farm sector for abandonment of its perceived moral obligations, and the extent of its special-interest lobbying have accelerated distrust of industrial-scale farming.[1] The result has been a shift in the basis for farm subsidy programs towards the promotion of sustainable or ethical farming, and 'clawing back' some of the special consideration that farming has enjoyed.

Shifting social attitudes

Chapter 5 suggests that, in Australia, we have seen the attitude of society shift from viewing irrigation farming as a strategically important contributor to social development goals of 'opening up the country', to a more sceptical stance that sees farming production as good, but irrigation water use as too frequently excessive. Particular media attention has been paid to the water required for cotton and rice, and of mega-enterprises with large-scale water rights, and to farmers who have conducted illegal activities such as unauthorised land clearing or water diversion. As in the USA, an attachment to the romantic historical image of small-scale farming remains, but the economics of this model are proving to be increasingly difficult. The social licence of irrigators to use water has become heavily challenged, and water use is now constrained through law to fit within the scientifically determined needs of the environment. The mechanisms to 'claw back' water for the environmental public good have not been unsympathetic to irrigators, even if the net impact has been to create significant challenges for the sector. In the last two decades, the creation (and then re-purchase by government) of secure water property rights has created new sources of wealth for 'water entrepreneurs', and irrigators have benefited from significant government investment in water research, infrastructure and efficiency improvements.

Two traditionally large-scale water users (cotton and rice) have embraced the need to achieve significant improvements in water use and have adopted purposeful and positive responses to shifting community attitudes. It is interesting to observe how these industrial-scale farming industries have responded to public criticism of their environmental impacts. Chapter 7 describes a cotton industry moving purposefully to change farming practices so as to reduce its impacts, and using transparent reporting to help to shift community views of its performance. The rice industry, through research and development and the Rice Environmental Champions Program, has embarked on a similar path of combining substantive change in farm practices with active communication to re-establish trust and defend its social licence. In both instances, the scale and sophistication of the industries has helped them to respond in a coherent manner, which has not always been the case with farming.

As in the USA, the media and activist non-government organisations have been influential, and continue to press for higher levels of resource efficiency, public accountability and social virtue from farmers. These demands are not only made in relation to water use. Biodiversity protection and animal welfare in particular, but also worker safety and welfare, continue to be fields in which the practices of farmers have been subject to much criticism, and lobbying for more stringent legal controls. At the same time, farming is being presented as a potential partial solution to Australia's net carbon emissions, requiring that farmers and their advocates convince a less-than-trustful public and government that farming can be a central part to the solution of an intractable environmental problem, rather than being perceived as part of the problem.

The nature of these challenges suggests that the farm sector will need to secure strong community and political support, backed by good science, if it is to advance its interests. The examples provided in this book suggest that creating a strong platform for this will require true leadership, including a willingness to embrace criticism that may be initially seen as unfair.

On the other side of the world from Australia, in Iceland, farmers have seen media exposure of the harmful effects of their practices lead to radical changes in government policies, notably with a shift in economic support away from production *per se* towards sustainable farming practices. This has been mirrored throughout the European Community, where public good outcomes of farming, notably ecological outcomes, have become the dominant concern for farming policy. In Europe in particular, we can point to a close relationship between community social and ecological concerns and the design of legal constraints or economic supports for agriculture. In addition to the shift in the basis of European Common Agricultural Policy payments towards sustainable farming practices outlined in Chapter 10, a further indication of the public concern for sustainability is the extent of the controls over the use of genetic modification technologies in Europe. GMO products, with the potential to improve farming profitability and productivity, have been held back through legislation, in response to the strength of community perceptions of risk. The difference in treatment of GMO issues in Europe compared with Australia and the USA is a clear reflection of how different political philosophies can result in different degrees of licence for farming (noting that, through the CAP, European farmers also enjoy significant payments for forms of farming that are perceived to be supportive of the public good, such as organic farming).

In developing countries such as India, China and Bhutan (Chapter 6), the social licence for farming is evolving to require greater social accountability alongside a push to modernise. As farming has become more industrialised and remote from rural communities, it has become more efficient and less important as an employer, and as communities have become legally empowered to demand social and ecological justice, farming has become subject to tighter legal controls and political pressure. Although food scarcity and the drive towards development combine with the relative economic importance of farming to ensure greater latitude to farmers in these countries to maintain practices that are unlikely to be accepted in more developed economies, this latitude is shrinking. This is more particularly the case where, as in some of the examples outlined in Chapter 6, the constitution of the country prioritises protection of the environment. The social licence of developing country farmers is subject to the same trends and processes of evolution as that of more advanced economies.

Winners and losers

The supremacy of the citizen is not only political it is also economic. Consumer freedom is embedded within our political and economic system, ranking alongside political freedom. Consumers exercise economic power through their purchasing choices, and they are free to apply whatever criteria they wish to their spending. With greater wealth and more choice has come an increasing emphasis upon attributes of the producer and the production method, in addition to attributes of the products themselves. The economic social licence can produce winners and losers. One of the substantial challenges for farming is to find ways of converting consumer preferences for farming that meets their social good expectations into a premium for those farmers who invest to meet these expectations.

The organic produce movement is a useful illustration. Organic product consumers are prepared to spend more, and expend more search costs, to obtain products that they believe are less harmful to the environment. A similar dynamic of consumer preference exists with consumers of 'fair trade' products, or those who purchase with the intention of reducing 'food miles', or to buy products from their own national producers. These initiatives depend upon meeting the need of the consumer in two ways: producing farm products that comply with the consumer's special expectations, and providing information about the characteristics of the product or the methods of production that allow the consumer to distinguish which products meet this need.

Chapters 4 and 7 indicate that some advantages can be secured through transparent communication of investment to improve the social and environmental performance of farming.

The experience of the cotton industry and the organic movement suggests, however, that to achieve any benefit does require a significant commitment to accept and respond to community criticisms and concerns. Although communications are important, the essence is to ensure that they are backed by tangible performance driven by a disciplined strategy. This is further supported by the examination of voluntary GRI reporting (Chapter 13). It is difficult to present meaningful social and environmental impact reporting without the clarity of objectives and purposefulness of action that comes from making hard decisions about exactly what impacts the industry is prepared to be accountable for, or what opportunities for improved social contribution it believes are feasible for it to make. Additional reporting without this coherence does seem to hold the potential to add materially to costs and complexity, with limited gains in terms of social licence.

Pro-sustainability values are by no means foreign to farming culture. A culture of stewardship exists among many farmers, as is well illustrated by the views discussed in Chapter 14. However, as in the rest of society, the views of farmers about the environment and social justice issues are not homogenous, and economic pressures materially increase the likelihood of unsustainable farming practices being used (even where the farmer may wish to adopt a higher standard of care). Movements such as Farmers Heal the Land, Landcare, organic farming and the many farmer-based sustainability initiatives all speak of the practical commitment of many farmers to high ethical principles. The absence of mechanisms to recognise higher standards of stewardship and contributions to the public good is an inhibitor to the further strengthening of such initiatives. The lack of tangible economic benefits for investment in delivery of public goods such as environmental services probably acts as an inhibitor to more widespread adoption of many farming practices that would be considered by the general public as superior to traditional approaches.

Farmers and other groups have attempted to respond to the need for objective consumer information about products and production processes in a number of ways, which are discussed in part in Chapters 4, 7 and 16. Farming EMS systems, reported standards and certification schemes, consumer brands and farming best practice systems have evolved in response to farmers' desires to improve their own sustainable farming practice, and in pursuit of economic recognition for better public good performance. Despite attempts by farming social entrepreneurs to achieve closer integration between farmer self-enforced standards and the requirements of legislation, and scholarly discussion that suggests the potential for economic and ecological gains from co-regulation, progress on this front has been grudging. As Chapter 2 illustrates, this may in part be due to distrust born of past actions of some farmers, and to caution on the part of government. Regardless of the cause, this chapter suggests that an insufficiency of trust is an impediment to co-regulatory reforms that could strengthen accountability while reducing administrative burdens on farmers.

Rights and responsibilities

Citizens and consumers enjoy extensive legal rights to protect their property, protect their liberty and exercise their freedoms. People who allow noise, smells or other harms to flow from their property, or infringe on their neighbours' property, can be restrained (except if they are subject to special protection, as with the US 'right to farm' laws). Companies who mislead consumers can be sued, as can other citizens who seek to damage the reputation of the citizen.

It is not only citizens whose interests are protected in contemporary democracies. In the public interest, protection is extended to the market itself (through trade practices and consumer legislation, or through laws ensuring free trade or to prevent abuse of market power), to the environment and to animals. In Australia alone, there are more than 200 laws to protect environmental values (Martin and Verbeek 2000), and the suite of public interest rules in all countries is extensive.

It is against the backdrop of the fundamental operations of democracy and the consumer fundamentals of capitalist economics that farmers' social licence, stewardship and public accountability issues should be considered. Farmers as citizens benefit from the variety of freedoms protected by Parliament and the law, but are accountable against the legal duties and social obligations of citizens. Because farmers control large parts of the natural landscape, rely on the use of resources (such as water or public lands) that they do not own, sometimes carry out activities that are harmful or offensive to neighbours, use animals and engage in trade, they are subject to many rules. As society raises its expectations and competition for markets or resources intensifies, the requirements placed on farmers are likely to escalate. If the farm sector is not able to demonstrate that it has in place effective voluntary programs that will reliably deliver the public goods that the community wants, the pressure for regulation is likely to intensify. In this context, producers who are able to earn the trust of the community and convert this to a 'brand advantage' ought be able to secure economic and political advantages.

Society can also decide to support, rather than restrict, farmer's activities, with 'right to farm' laws or supports for production or conservation activities, or to facilitate change by providing education, extension and financial payments. Consumers can choose to pay a premium for products that they consider best meet their social, as well as consumption, needs, or administrators can determine that it is in the public interest for discretion to be exercised in the farmers' favour.

Although farmers pride themselves on being masters of their own destinies, the modern reality is that this is increasingly a form of negotiated autonomy. The case studies in this book suggest that the hegemony of the majority is a reality with which farming has to deal effectively. Creating or defending a strong social licence to farm is not a concern that is likely to be successfully managed by defensive legal actions, marches on Parliament, corporate communications or media messages about the importance of food security. These tactical actions may assist at the margins, but ultimately the power of the vote and the dollar lies in the hand of the sceptical majority. The case studies from the Australian cotton and organic produce industries, and the Icelandic sheep farming industry, suggest that the likely path to success in coping with social licence issues is to seek a true alignment between the actions of farmers and the expectations and preferences of society, difficult though this may be to achieve.

Strategies, tactics and instruments to defend social licence

The discussion above suggests that the social licence is a reflection of trust between society and farmers as a group, and that this trust can be easily lost, but not easily regained. For farmers, losing part of their historical status as trusted stewards of natural resources is of more than emotional or social significance. This status stands behind economically important political and administrative decisions about financial and legal supports for farming. It is material to how the political system responds to other stakeholders who demand increased legal controls over farming. The examples included in this book demonstrate that if farmers as a group are not trusted to meet the public good expectations of the general community, then society's response is to control their behaviour through regulation, or to restrict their access to economic support or natural resources.

Is it likely that pressures on the social licence of farmers will diminish? Although it is likely that food and fibre production will become more important with projected scarcities, there is little reason to believe that this will translate into liberalisation of the legal or economic pressures on farming to demonstrate higher standards of social and ecological responsibility.

Political demands to protect natural resources seem set to continue, simply because of the demands upon these resources. Awareness of declines in biodiversity, water availability and the capacity of natural systems to absorb pollutants are growing, and climate change will not reduce these concerns. Farming should expect to be required to continually demonstrate its good stewardship.

How then can farmers best defend their social licence? Will it be effective to apply contemporary approaches such as industry self-regulatory standards, voluntary reporting schemes, political lobbying and public relations, or legal actions to defend property rights? Will it be sufficient for farmers to meet ever-increasing legal requirements, to ensure that society values the stewardship efforts of farmers? Do the strategies, tactics and instruments farmers use to defend the social licence (or governments to force compliance with social expectations) suffice or will something more radical be needed?

Chapter 3 suggests that, although formal legal requirements require compliance with specific rules, social and political process seek virtuousness in how farmers act, which goes beyond these formal duties. What is virtuous, and indeed how virtuous one can afford to be in a highly competitive commercial world, are questions that are difficult to answer. This is particularly so because society's expectations are far from unified and are constantly changing. Conducting farming in ways that speak to society of the virtue of farmers is increasingly difficult as production becomes more industrialised, and as margins for food and fibre commodities shrink in response to the capacity of larger scale and more technology intensive farming to produce at very low cost.

The demand for virtuousness, and the ambiguity and diversity of perceptions of what that means, drives the expansion of formal legal requirements with which farmers are expected to comply. This is illustrated with the creation of duties of care for the environment (Chapter 14) that formalise the ambiguous social expectations of care for the environment in the law. Chapter 14 indicates that this is not likely to remove uncertainty and the potential for conflict over farmer stewardship expectations of the community, nor to prevent further legal responsibilities being created. Legal formalisation converts a social and political issue into a matter to be determined by administrators, and eventually by the courts, but, even when the meaning of 'duty of care' is legally determined, this will not prevent society seeking further guarantees of social accountability from farmers.

Engaging positively with the social licence challenge for farming seems to be the most sensible strategy for most farming sectors. Eventually and inevitably, consumer preference and political power are likely to prevail over sectoral interests, and so, although defensive strategies may be useful in the short term, they seem unlikely to be the most effective in the longer term. The positive examples in this book have shown that farmers can take effective steps to renegotiate their social licence. Successfully negotiating with a powerful counterparty usually requires that you take their needs and preferences seriously, to find creative ways of making '2 plus 2 = 5'. The first step is to listen with a view to meeting their requirements to the maximum extent possible. This is difficult to do while denying the legitimacy of the other party's expectations, or if you are committed to avoidance of change in your own practices. Examples from the USA, Europe, Iceland and Australia suggest that it is possible to renegotiate the supports and the restrictions that surround farming. The positive examples have all demonstrated that a willingness to embrace change, even where doing so is costly and difficult, is a precondition of this renegotiation.

Can farmers adapt to changing social expectations, while securing freedom and support for their enterprises? Where farmers have been effective in embracing increased accountability, this has required a hard-nosed strategy that includes implementation of transformational change. This has required great leadership. It has also required sophisticated thinking, to target with care the issues against which the industry are prepared to be held accountable for, or the opportunities where it can demonstrate an increased public good contribution.

Although we have not set out to provide a strategic framework for farming to respond to the social licence issues it faces, some of the chapters indicate directions that might be fruitful for the farm sector in pursuing this end. Chapter 11 highlights that the European Community has been able to create a principles-based approach to determining when it is reasonable to impose new constraints on farming, which could be relevant in setting clearer principles for government intervention. Chapter 16 suggests a conceptual design for an integrated and streamlined co-regulatory arrangement to marry regulation with voluntary management and better consumer information. This aligns with other discussions of the need for (and potential design of) streamlined natural resource governance arrangements for the farm sector. Chapter 12 outlines a framework for determining a strategic approach to prioritising the social and ecological issues that a strategically focused corporate social responsibility program ought to consider. Such guidance, coupled with the practical experience of farm organisations that have undertaken the challenge of renegotiating their social licence, provides a starting point for developing a coherent response to the challenges that have been highlighted in this book.

As discussed at the outset of this book, it would be unwise to trivialise what responding to changed community expectations means to farmers. Effective transformation involves significant investment, and perhaps sacrificed income. Transformational change requires heroic leadership, and integrity in the face of difficult choices and conflict. Although change offers benefits to the larger community, the costs will often be borne by responsible farmers more than any other group. This book does, however, demonstrate that, although change may be painful and costly in the short term, embracing the social licence challenge in a positive manner may be the most effective way to ensure continued support from the community.

Endnotes

1 There are many such online exposes; for example, 'Unleashed' from Animals Australia; 'Mike King experiences the horrors of a NZ pig farm' from Open Rescue New Zealand; or the recently widely promoted 'Food, Inc.'

References

Martin P (2010) Finding a partial cure for rural policy insanity. *Australian Farm Policy Journal* **7**, 2.

Martin P and Verbeek M (2000) *A Cartography for Natural Resource Law: Finding New Paths to Effective Resource Regulation*. Land and Water Australia, Canberra.

Index